for GCSE mathematics

Practice for Foundation

PUBLISHED BY THE PRESS SYNDICATE OF THE UNIVERSITY OF CAMBRIDGE
The Pitt Building, Trumpington Street, Cambridge, United Kingdom

CAMBRIDGE UNIVERSITY PRESS
The Edinburgh Building, Cambridge CB2 2RU, UK
40 West 20th Street, New York, NY 10011-4211, USA
477 Williamstown Road, Port Melbourne, VIC 3207, Australia
Ruiz de Alarcón 13, 28014 Madrid, Spain
Dock House, The Waterfront, Cape Town 8001, South Africa

http://www.cambridge.org/

© The School Mathematics Project 2003
First published 2003

Printed in the United Kingdom at the University Press, Cambridge

Typeface Minion   *System* QuarkXPress®

A catalogue record for this book is available from the British Library

ISBN 0 521 89032 2 paperback

Typesetting and technical illustrations by The School Mathematics Project
Illustrations on pages 18, 20, 36, 51, 52, 55, 78, 85–87,
118, 119, 135 and 169 by Chris Evans
Illustrations on page 50 by David Parkins
Photograph on page 112 by Paul Scruton
Cover image © Getty Images/Nick Koudis
Cover design by Angela Ashton

Contents

1 Rooms

Section A

1 Find the area of these rectangles in cm².

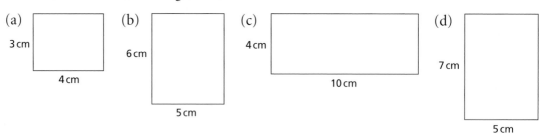

(a) 3 cm, 4 cm
(b) 6 cm, 5 cm
(c) 4 cm, 10 cm
(d) 7 cm, 5 cm

2 Find the area of these rectangles in cm².

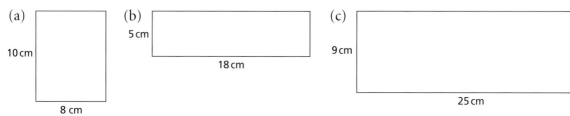

(a) 10 cm, 8 cm
(b) 5 cm, 18 cm
(c) 9 cm, 25 cm

3 (a) What is the area of this tile?

5 cm
10 cm

   (b) What wall area could you cover with 20 tiles?
   (c) What wall area could you cover with 200 tiles?

4 Find the area of these rectangles.

(a) 20 cm, 55 cm
(b) 22 cm, 32 cm

5 Rehana has some tiles that measure 5 cm by 5 cm.
   How many does she need to tile an area of 80 cm by 160 cm?

Section B

This is the plan of the upstairs of a house.

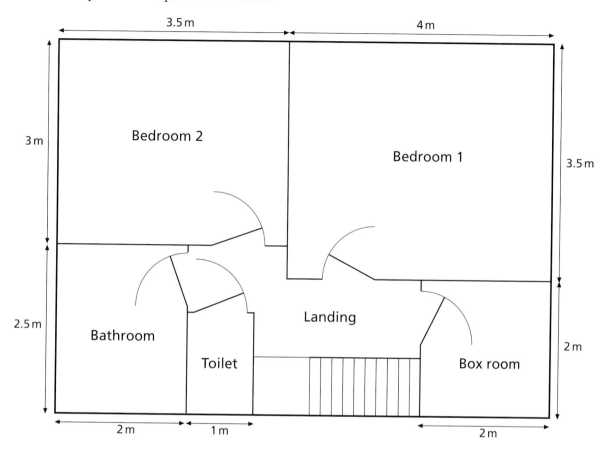

1 Find the area of each of these.

   (a) The box room   (b) The bathroom   (c) Bedroom 1   (d) Bedroom 2

2 What, roughly, is the area of the toilet?

3 Carpet costs £18 per square metre.
   How much will it cost to carpet each of these?

   (a) The box room   (b) Bedroom 1

4 A one-litre can of floor stain covers $2\,m^2$ and costs £4.00.

   (a) How many cans of floor stain would you need to buy for the floor
       in Bedroom 2?

   (b) How much would the floor stain for Bedroom 2 cost?

Section C

Remember:
- Sketch the shape.
- Split it into rectangles. or
- Write in any measurements you need.

1 Work out the area of each of these floors.

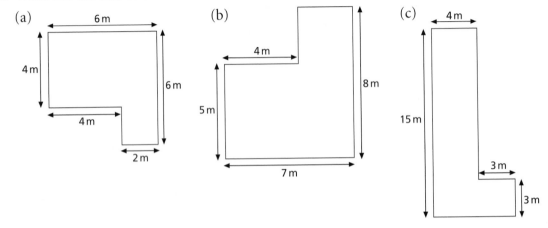

(a) 6 m 4 m 6 m 4 m 2 m

(b) 4 m 8 m 5 m 7 m

(c) 4 m 15 m 3 m 3 m

2 Work out the area of each floor.

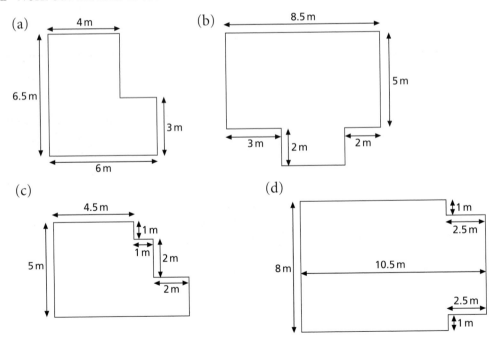

(a) 4 m 6.5 m 3 m 6 m

(b) 8.5 m 5 m 3 m 2 m 2 m

(c) 4.5 m 1 m 1 m 2 m 5 m 2 m

(d) 1 m 2.5 m 8 m 10.5 m 2.5 m 1 m

3 Find the perimeter of each floor in question 1.

Section D

1 Write each of these measurements as a decimal of a metre.
   (a) 60 cm   (b) 42 cm   (c) 7 cm   (d) 98 cm
   (e) 1 m 12 cm   (f) 1 m 73 cm   (g) 2 m 5 cm   (h) 5 m 8 cm
   (i) 87 cm   (j) 123 cm   (k) 402 cm   (l) 12 m 4 cm

2 Change these measurements into centimetres.
   (a) 0.35 m   (b) 0.2 m   (c) 0.8 m   (d) 1.9 m
   (e) 4.32 m   (f) 2.05 m   (g) 0.07 m   (h) 4.3 m

3 By changing measurements into decimals of metres, find
   the area of each of these rooms in m².

   (a) 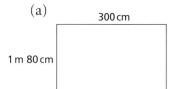   (b)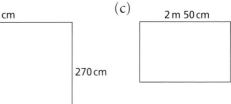

   (c)

4 Find the area of these rugs in cm².
   (a) 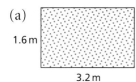   (b) 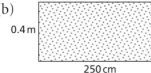   (c)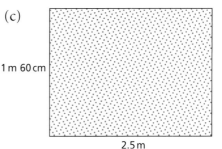

5 Each rug in question 4 needs carpet tape around the edge.
   What length of tape is needed for each, in centimetres?

6 Write each length of tape in question 5 in metres.

2 Weather watching

Section A

This data shows the highest temperatures recorded at Banff and Vancouver in Canada for a week in January.

Highest temperature recorded (°C) in January							
Date	3rd	4th	5th	6th	7th	8th	9th
Vancouver	5	3	7	4	8	2	3
Banff	⁻7	⁻3	⁻8	⁻8	⁻5	⁻9	⁻7

1 What was the highest temperature at Banff on January 5th?

2 What was the highest temperature at Vancouver on January 8th?

3 What was the difference between the highest temperatures in Banff and Vancouver on

   (a) 4th January   (b) 7th January

4 On which date was the difference between the highest temperatures in Banff and Vancouver the most?

   What was that difference?

5 On 10th January the difference between the temperatures was 9°C. In Banff it was ⁻5°C.

   What was the temperature in Vancouver?

Section B

This graph shows the average temperatures recorded for one year in Thule (Canada) and Cape Town (South Africa).

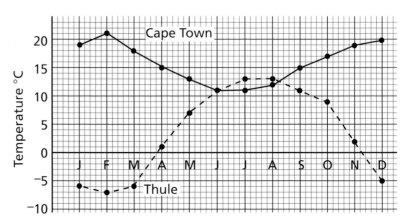

1 What was the average temperature in Thule in March?

2 In what months was the average temperature higher in Thule than in Cape Town?

3 What was the difference between the temperatures in Thule and Cape Town

   (a) in April   (b) in June

4 When was the difference between the temperatures the most?

These tables show the average monthly rainfall for one year in Rome (Italy) and Irkutsk (Russia).

Irkutsk Average rainfall (mm)	
January	0
February	0
March	0
April	1
May	3
June	7
July	10
August	8
September	4
October	2
November	1
December	0

Rome Average rainfall (mm)	
January	6
February	5
March	3
April	4
May	5
June	2
July	1
August	2
September	6
October	9
November	8
December	7

5 Draw a line graph to show the data for Rome and Irkutsk.
(Do both graphs on the same pair of axes.)

6 Use your graph to answer these questions.

  (a) Which place is wetter in June?

  (b) For how many months was it wetter in Irkutsk than Rome?

For each month, this chart shows the average number of hours a day of sunshine in Kendal.

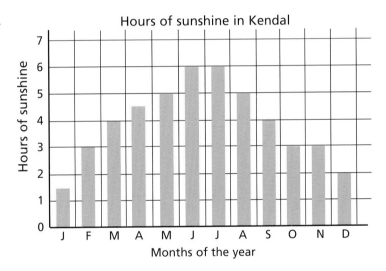

7 How many hours a day of sunshine were there on average in these months?

  (a) January   (b) September

8 In which months were there more than 4 hours a day of sunshine on average?

9 Draw a bar chart for the data in this table.

Average daily hours of sunshine in Cambridge											
J	F	M	A	M	J	J	A	S	O	N	D
2	3	2	5	5	6	6	5	4	3	2	2

Section C

This data shows Tim's records of the weather each day in January in York.

Weather watch: January
bright, cloudy, snow, cloudy, rain, bright, rain, cloudy, bright, cloudy, snow, rain, cloudy, snow, cloudy, snow, snow, rain, snow, rain, snow, rain, snow, cloudy, rain, snow, snow, rain, snow, rain, rain

1 (a) Copy and complete this frequency table.

Weather watch: January		
	Tally	Frequency
Bright		
Cloudy		
Rain		
Snow		

(b) Copy and complete this bar chart on squared paper.

(c) What was the modal type of weather in January?

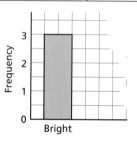

2 This data shows the hours of sunshine in York each day during July.

Hours of sunshine: July
6, 4, 5, 6, 7, 3, 4, 6, 5, 2, 6, 1, 5, 3, 7, 5, 4, 6, 8, 5, 3, 5, 4, 2, 1, 3, 5, 6, 8, 2, 4

(a) Copy and complete this frequency table.

Hours of sunshine		
Hours	Tally	Frequency
1		
2		
3		
4		
5		

(b) Copy and complete this bar chart.

(c) What was the frequency of 3 hours of sunshine?

(d) What was the mode of the number of hours of sunshine?

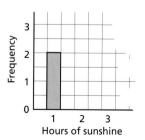

Section D

1 Find the mean and range of these sets of weather data.

(a) Oktas of cloud in one week: 1 6 5 3 0 1 5

(b) The hours of sunshine at 6 resorts on one day: 5 5 7 8 5 3

(c) Daily rainfall in mm at 8 resorts: 2.8 1.2 1.5 0.8 1.2 1.4 1.0 2.1

(d) The midday temperature in °C at some resorts:

23 18 25 20 19 22 26 20 15 21

(e) The number of days on which it rained in some resorts in August:

11 15 8 6 5 8 7 6

(f) The temperature in °C at midday at some ski resorts:

5 2 6 5 6 4

2 The weather for various places in Scotland is given below for a day in August and a day in September.

August 17th 2003	Sun (hrs)	Rain (mm)	Temp (°C)
Aberdeen	0.5	11	16
Aviemore	1.6	5	17
Edinburgh	7.9	0	18
Eskdalemuir	4.4	1	15
Fair Isle	11.7	0	14
Kinloss	0.7	2	16
Lerwick	6.4	3	15
Leuchars	6.5	1	19
Stornoway	0.7	2	15
Tiree	3.6	8	15

September 1st 2003	Sun (hrs)	Rain (mm)	Temp (°C)
Aberdeen	1.3	0	21
Aviemore	2.1	0	19
Edinburgh	2.7	0	20
Eskdalemuir	0	1	17
Fair Isle	0	4	15
Kinloss	1.6	0	19
Lerwick	2.9	0	14
Leuchars	2.5	0	21
Stornoway	1.3	1	15
Tiree	0.1	5	16

(a) Find the mean and range of the amount of rain that fell on

(i) August 17th   (ii) September 1st

(b) Which day had the most rain on average in these Scottish places?

(c) Find the mean and range of the temperature for each of the days.

(d) Which of the two days had the highest temperature on average?

(e) Which day had the most hours of sun on average in these Scottish places?

3 Reversing the flow

Section A

1 Copy and complete each of these mathematical whispers.

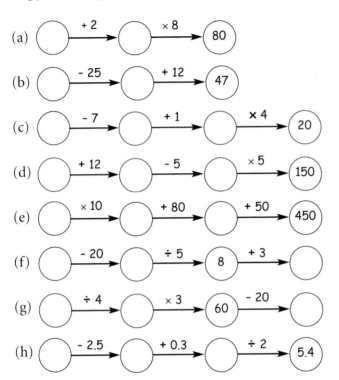

2 Draw an arrow diagram for each of these whispers.

For each one work out the number Emily started with, using a calculator if necessary.

(a) Emily thinks of a number and adds 2.6.
Allen takes Emily's answer and multiplies by 3.5.
Alex divides Allen's answer by 3 and ends up with 7.

(b) Emily thinks of a number and subtracts 8.4.
Allen takes Emily's answer and divides by 2.
Alex adds 4.2 to Allen's answer and ends up with 10.

(c) Emily thinks of a number and divides by 1.5.
Allen takes Emily's answer and multiplies by 4.5.
Alex divides Allen's answer by 1.8 and ends up with 100.

Section B

1 Suppose the letter x goes into each of these puzzles.

Match each puzzle with one of the equations underneath.

(a)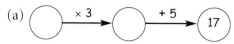

(b)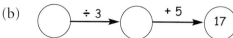

(c)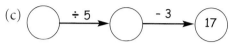

(d)

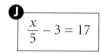

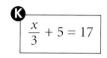

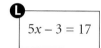

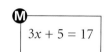

2 Draw a puzzle for each of these equations.

(a) $3x + 5 = 14$

(b) $\dfrac{x}{3} - 4 = 6$

(c) $4x - 8 = 36$

(d) $10x + 6 = 46$

(e) $\dfrac{x}{4} + 4 = 7$

(f) $\dfrac{x}{5} + 3 = 5$

3 Match this equation to one of the puzzles below: $\quad 7r - 3 = 25$

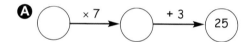

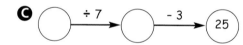

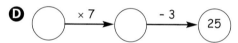

4 Suppose the letter p goes into each of these puzzles.

Write down an equation for each puzzle.

(a)

(b)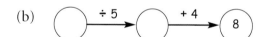

(c)

(d)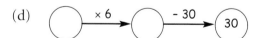

(e)

(f)

(g)

(h)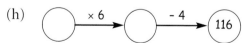

Section C

Try not to use a calculator for these questions.

1 Copy the arrow diagram for each of these equations.
Then reverse the diagram to solve the equation.

Check your solution.

(a) $2k + 4 = 8$

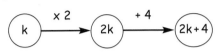

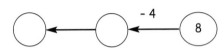

(b) $7b - 2 = 47$

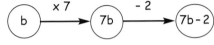

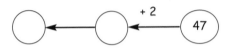

(c) $\frac{r}{5} - 2 = 2$

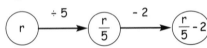

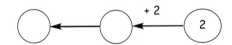

(d) $\frac{z}{3} - 8 = 19$

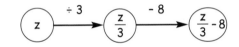

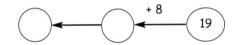

2 Draw an arrow diagram for each of these equations.
Reverse the diagram to solve the equation.

Check each of your solutions.

(a) $3w + 2 = 23$

(b) $2d - 10 = 10$

(c) $\frac{s}{8} + 9 = 10$

(d) $5k - 9 = 16$

(e) $5b + 2 = 27$

(f) $\frac{a}{3} - 2 = 0$

(g) $\frac{m}{8} + 4 = 5$

(h) $5z + 6 = 51$

(i) $7e - 7 = 7$

(j) $\frac{y}{2} + 10 = 14$

You may need to use a calculator for the next question.

3 Solve each of these equations.

(a) $5b - 7 = 8.5$

(b) $\frac{c}{9} - 1.3 = 4.7$

(c) $4n - 9.8 = 13$

(d) $8f - 6.4 = 33.6$

(e) $\frac{g}{5} + 9.4 = 15$

(f) $\frac{p}{1.5} + 6.2 = 20.2$

Section D

1 Write down an equation for each of these number puzzles.

   Solve your equation using arrow diagrams.

(a)
```
I think of a number.
 •  I multiply by 5.
 •  I add 6.
My answer is 51.
What was my number?
```

(b)
```
I think of a number.
 •  I divide by 7.
 •  I add 18.
My answer is 20.
What was my number?
```

(c)
```
I think of a number.
 •  I multiply by 2.
 •  I subtract 9.
My answer is 11.
What was my number?
```

(d)
```
I think of a number.
 •  I multiply by 9.
 •  I subtract 5.
My answer is 22.
What was my number?
```

(e)
```
I think of a number.
 •  I divide by 2.
 •  I subtract 14.
My answer is 1.
What was my number?
```

(f)
```
I think of a number.
 •  I multiply by 4.
 •  I add 18.
My answer is 30.
What was my number?
```

(g)
```
I think of a number.
 •  I multiply by 4.
 •  I subtract 5.1.
My answer is 7.7.
What was my number?
```

(h)
```
I think of a number.
 •  I divide by 8.
 •  I add 9.5.
My answer is 14.
What was my number?
```

2 Write down a number puzzle for each of these equations.
   Solve the equation using an arrow diagram.

   (a) $4n - 2 = 18$

   (b) $\frac{n}{3} - 5 = 9$

   (c) $\frac{n}{5} + 2 = 4.8$

   (d) $8n - 1 = 47$

   (e) $8n + 15 = 71$

   (f) $\frac{n}{10} + 1.4 = 2$

3 Write down a number puzzle with the answer 12.
   What was your starting number?

4 Write down a number puzzle that starts with the number 6 and
   finishes with the answer 20.

4 *Mental methods*

Section A

1 In the number **372 984**, the **9** stands for 9 hundreds.
   (a) What does the figure **2** stand for?   (b) What does the figure **7** stand for?

2 Work out
   (a) 2615 + 100   (b) 33 961 − 1000   (c) 12 574 + 300   (d) 972 513 + 2000

3 The mileometer on Sarah's car reads **3 8 2 7 9**.
   What will it say when she has gone these distances?
   (a) 20 miles   (b) 500 miles   (c) 3000 miles   (d) 20 000 miles

4 Write these numbers in order, smallest first.
      35 675   36 143   36 002   35 432   35 141

Section B

1 (a) Round 43 761 to the nearest hundred. (b) Round 461 958 to the nearest thousand.
   (c) Round 647 208 to the nearest ten thousand.

2 The second highest mountain in the world is K2 at 28 251 feet.
   Round the number 28 251 to the nearest
   (a) thousand   (b) ten   (c) hundred   (d) ten thousand

3 Round each of these numbers to one significant figure.
   (a) 3756   (b) 42 785   (c) 119 253   (d) 70 236   (e) 684

4 Below are the areas, in square kilometres, of some islands.
   Round these areas to one significant figure.
   Greenland 2 175 600 km^2 Java 126 602 km^2 Sri Lanka 65 610 km^2

Sections C and D

1. (a) 13.7×10 (b) 0.48×100 (c) 10×3.76 (d) $150.3 \div 10$ (e) $231 \div 100$
   (f) 6.2×1000 (g) $5.72 \div 100$ (h) 100×0.062 (i) $58.3 \div 100$ (j) $474 \div 1000$

2. (a) 140×20 (b) 22×30 (c) 213×200 (d) 1.5×400 (e) 21×600
   (f) 110×50 (g) 12×40 (h) 34×2000 (i) 8×400 (j) 102×30

3. (a) 40×70 (b) 30×60 (c) 200×30 (d) 400×200 (e) 60×500
   (f) 800×30 (g) 40×400 (h) 30×2000 (i) 80×70 (j) 700×90

4. Work out the cost, in pounds, of
   (a) 80 crocus bulbs at 30p each   (b) 400 daffodil bulbs at 20p each
   (c) 500 lily bulbs at 40p each   (d) 7000 bluebell bulbs at 20p each

5. A large concert hall has 8000 seats.
   It sells tickets at £30 each.

   How much money does it make if all the tickets are sold?

Section E

1. Work out a rough estimate for each of these.
   (a) 28×42 (b) 28×48 (c) 283×22 (d) 57×82 (e) 29×624

2. Sethi rents a workshop. It costs £47 a week.
   Estimate how much she pays in a year (52 weeks).

3. Jacqui's garden is rectangular.
   It is 38 metres long and 31 metres wide.

   Estimate its area.

4. A milkman delivers around 480 pints of milk a day.
   He delivers 361 days a year.

   Roughly how many pints of milk does he deliver in a year?

5 | *Made to measure*

Section A

1 Measure the lengths of these feathers.
   Write down their measurements in mm and then in cm.

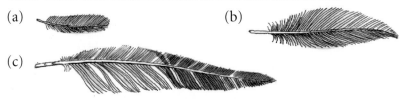

   (a)   (b)

   (c)

2 This table shows the lengths of some small birds.
   Copy the table and complete it.

Firecrest

Bird	Length (mm)	Length (cm)
Goldcrest	90	
Firecrest		8.5
Pied flycatcher		12.5
Blue tit	113	
Long-tailed tit		14

3 (a) Which of the birds in the table above is the shortest?

   (b) Which is the longest?

4 | 5.5 cm | | 6 mm | | 3 cm | | 2.5 cm | | 52 mm |

   Write the lengths above in order of size, smallest first.

5 This table shows the wingspans of some birds of prey.
   Copy the table and complete it.

Bird	Wingspan (cm)	Wingspan (m)
Buzzard	125	1.25
Red kite	195	
Golden eagle		2.2
Osprey	170	
Kestrel	80	

Red kite

6 | Gannet 1.6 m | | Fulmar 102 cm | | Cormorant 1.35 m | | Green cormorant 90 cm |

   (a) Which of the sea birds above has the largest wingspan?

   (b) Which has the smallest wingspan?

7 Write these lengths in cm.

   (a) 3 m   (b) 4.5 m   (c) 6.45 m   (d) 0.5 m   (e) 0.68 m

Section B

1. How many grams are there in each of these?

   (a) 3 kg   (b) 1.5 kg   (c) 0.6 kg   (d) 3.2 kg   (e) 14 kg

2. Write these in kg.

   (a) 5000 g   (b) 1800 g   (c) 16 000 g   (d) 4250 g   (e) 600 g

3. Copy and complete this table. It shows the weights of some birds in grams and in kilograms.

   (a) Which of the birds in the table is the heaviest?

   (b) Which is the lightest?

Bird	Weight (g)	Weight (kg)
Herring gull	1200	
Common gull	450	
Great black-backed gull	2000	
Mute swan		12
Canada goose		4.5

4. How many metres are there in each of these?

   (a) 4 km   (b) 1.8 km   (c) 2.25 km   (d) 0.5 km   (e) 20 km

5. Write these in km.

   (a) 3000 m   (b) 6500 m   (c) 15 000 m   (d) 850 m   (e) 1250 m

6. This table shows the distances some pupils walk to school, in m and in km.

   Copy the table and complete it.

   (a) Which of the pupils lives furthest from the school?

   (b) Which lives closest to the school?

	Distance (m)	Distance (km)
Anna	1200	
Bethan		1.9
Ron	500	
Shamraz	650	
James		0.3

7. Petra planned to jog one kilometre each day. How many times would she need to jog

   (a) along a 100 m running track?

   (b) around a 250 m circuit?

   (c) around a 200 m circuit?

Section C

1 Write these measurements in metres.

  (a) 3000 mm   (b) 2500 mm   (c) 3260 mm

  (d) 15 000 mm   (e) 600 mm   (f) 75 mm

2 How many millimetres are there in each of these?

  (a) 2 m   (b) 30 m   (c) 1.7 m   (d) 0.8 m   (e) 0.04 m

3 Susan's bedroom wall is 3 metres long.
  Can she fit all of these units along the wall?

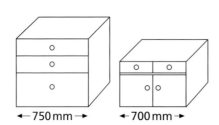 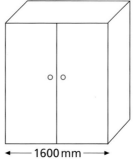

←750 mm→   ←700 mm→   ←——1600 mm——→

4 Write these lengths in order of size, smallest first.

  1.9 m   250 mm   1670 mm   0.45 m   0.8 m

5 How many millimetres are there in one kilometre?

6 Copy and complete this table showing how much some bottles hold,
  in litres and in millilitres.

Capacity (litre)	Capacity (ml)
2	
1.4	
	1200
	750
	50

7 Kate has a 2 litre bottle of cola.
  How many 200 ml glasses can she fill?

8 How many 250 ml glasses can be filled from each of these bottles?

  (a) 1 litre   (b) 0.75 litres   (c) 1.5 litres   (d) 3 litres.

9 Write these quantities in order of size, starting with the smallest.

    5 g   400 mg   2.5 g   1600 mg   250 mg

7 Decimal places

Section A

1 What number does each arrow point to?

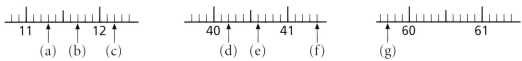

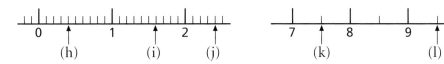

2 Which of these numbers are between 3.7 and 5.3?

3.5, 4, 3.9, 5.4, 1.6, 4.8, 6.1, 5

3 Put these numbers in order, smallest first.

5.3, 3, 4.8, 2.7, 8.5, 0.7, 6.3, 0.3

4 Put these numbers in order, largest first.

0.9, 8.2, 2.7, 4, 0.4, 3.6, 2.1, 0.8

5 The numbers in this pattern go up by 0.3 each time.

0.3, 0.6, 0.9, …

What are the next three numbers in the pattern?

6 Here are the heights of 6 people.

  Pam 1.6 m   Charlie 1.6 m   David 1.3 m   Hayley 0.9 m   Jackie 1.7 m   Harry 1.5 m

(a) Who is the tallest?

(b) Who is the shortest?

(c) What is the difference in height between the tallest and shortest person?

7 What number does each arrow point to?

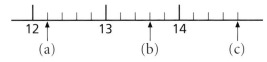

Sections C and D

1 What number does each arrow point to?

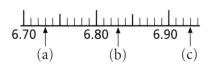
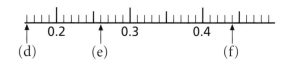

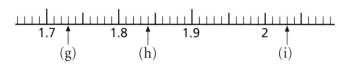

2 Put these cards in order from the smallest number to the largest.
What word do you get?

1.85	0.07	1.83	0.16	1.75	1.8
R	B	E	U	R	G

3 Which of these numbers are between 3.28 and 4.5?

4.8, 3.3, 2.57, 3.8, 4.25, 4.03, 3.07

4 Put these numbers in order, smallest first.

6, 6.9, 6.35, 6.09, 5.89, 5.6, 5.98

5 Put these heights in order, tallest first.

1.83 m, 1.57 m, 1.63 m, 1.85 m, 1.68 m, 1.75 m

6 A parcel weighs 1.78 kg.
Which of these weights is the closest?

1.07 kg   1.75 kg   1.8 kg

1.83 kg   1.74 kg

7 A hall is 2.3 metres wide.
These are the widths of some carpets.

Which ones are too wide to fit in the hall?

8 What number is halfway between each of these?

(a) 2.3 and 2.8   (b) 3.46 and 3.48

(c) 1.8 and 1.9   (d) 5.89 and 5.93

Symmetrical shapes

Section A

1 Describe fully each of the shaded shapes below.

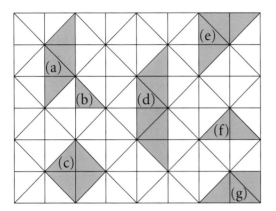 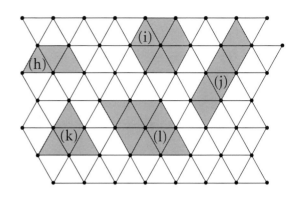

2 This pattern is made
from a regular octagon.
The shaded triangle has
corners at C, D and J.

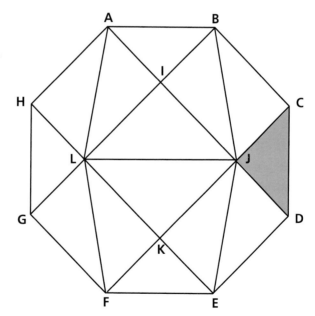

(a) What special type of triangle is the shaded triangle?

(b) Describe the shape with corners A, B, J and L.

(c) Describe the shape with corners B, J, E and L.

(d) List the corners of a shape which is a parallelogram.

(e) List the corners of a triangle which is scalene.

Sections B and C

1 Here are some shapes drawn on triangular dotty paper.

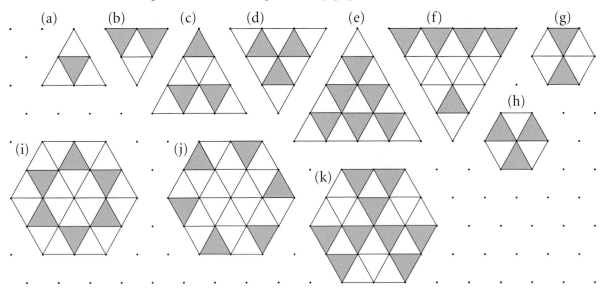

(a) (b) (c) (d) (e) (f) (g) (h) (i) (j) (k)

   (i)   Copy each of these shapes onto triangular dotty paper.

   (ii) Draw in the lines of reflection symmetry.

   (iii) Write the order of rotation symmetry underneath each diagram.

2 What sort of triangle is being described?

   (a)
> I have 2 equal sides.
> I have one line of symmetry.

   (b)
> I have no sides which are equal.
> I have no reflection symmetry.

   (c)
> I have 3 equal sides. I have 3
> lines of reflection symmetry.

3 Copy and complete the following.

   (a) *What am I?*
      I have six equal sides.
      I am a regular ....
      I have ..... lines of reflection symmetry.
      I have order of rotation symmetry .....

   (b) *What am I?*
      I have four sides.
      My opposite sides are parallel.
      I have no lines of symmetry.
      I am a ....
      I have order of rotation symmetry .....

Section D

1 All the shapes in this diagram are one half of a quadrilateral.
Each quadrilateral has line M as a line of symmetry.

For each quadrilateral

(i) give the coordinates of the point or points needed to complete the shape

(ii) write down the name of the quadrilateral

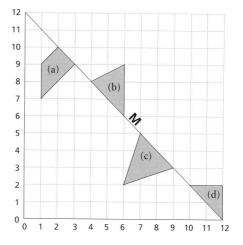

2 Draw a coordinate grid like the one above.
Draw a line going from $(0, 6)$ to $(12, 6)$. Label it 'M'.
Do these things for each of the sets of coordinates below.

  • Plot the points on the grid to make half of a shape.

  • For each shape, line M is a line of symmetry.
    Use this to draw the other half of the shape.

  • Write down the coordinates of the points you have added.

  • Describe the shape.

(a) $(1, 6)$ $(1, 12)$ $(3, 9)$ $(3, 6)$

(b) $(4, 6)$ $(5, 5)$ $(6, 6)$

(c) $(7, 6)$ $(8, 7)$ $(10, 7)$ $(11, 6)$

3 This diagram shows half of a shape which has rotation symmetry order 4 about centre $(0, 0)$.

(a) Give the point or points needed to complete the shape.

(b) What is the name of the completed shape?

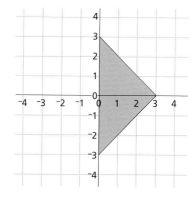

4 The points below make parts of shapes with rotation symmetry about the centre $(0, 0)$.
For each part, do this.

  • Draw a grid the same size as in question 3.

  • Plot the points you are given.

  • Complete the shape so it has the stated order of symmetry.

  • Give the coordinates of points you have added.

(a) $(^-3, 0)$ $(2, 2)$ $(3, 0)$ order 2   (b) $(^-4, 0)$ $(^-1, 2)$ $(1, 2)$ $(4, 0)$ order 2

(c) $(^-2, 0)$ $(^-2, 1)$ $(^-1, 2)$ $(1, 2)$ $(2, 1)$ $(2, 0)$ order 4

9 Working with formulas

Section B

1 Jason designs ornamental paving.

This is one of his paving designs.
With one rectangular block, he uses 6 square blocks.

With 2 rectangles,
he uses 8 squares.   Here is his design
for 3 rectangles.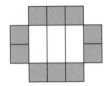

(a) How many squares does Jason use with 3 rectangles?

(b) How many squares does Jason need to make a design with 4 rectangles?

(c) How many does he need with 10 rectangles?

(d) How many does he need with 100 rectangles?

(e) Copy and complete this table.

Number of rectangles	1	2	3	4	5	6	10	100
Number of squares	6	8						

(f) Here are some rules connecting the *number of rectangles* and the *number of squares*.
Which of the rules is correct?

(number of squares = number of rectangles × 2 + 4)

(number of squares = number of rectangles + 2)

(number of squares = number of rectangles × 6)

(number of squares = (number of rectangles + 4) × 2)

(g) Suppose r stands for the *number of rectangles* and s stands for the *number of squares*.
Look at these formulas connecting s and r.

$s = r + 2$ $s = 2(r + 4)$ $s = 6r$

$s = 2r + 4$

Which of the formulas is correct?

2 Jason uses large white slabs and small grey blocks for this paving design.

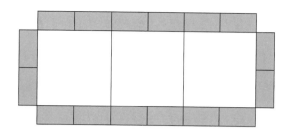

(a) How many grey blocks does Jason need to make this design using 10 white slabs?

(b) How many grey blocks does Jason need to make this design using 100 white slabs?

(c) Write a rule in words telling you the *number of grey blocks* you need when you know the *number of white slabs*.

Start your rule like this: *number of grey blocks = ...*

(d) Write your rule as a formula.
Use *w* to stand for the *number of white slabs* and *g* to stand for the *number of grey blocks*.

3 This paving design uses hexagons and triangles.

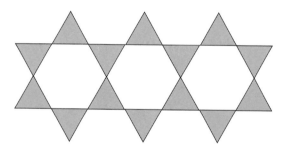

(a) How many triangles does this design need for 100 hexagons?

(b) For this design, work out a rule in words telling you the *number of triangles* you need when you know the *number of hexagons*.

(c) Write your rule as a formula.
Use *h* to stand for the *number of hexagons* and *t* to stand for the *number of triangles*.

Section C

1 Jason uses different designs of block paving.
   This is his *Basic* design.

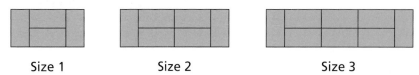

    Size 1   Size 2   Size 3

   (a) How many blocks does Jason use in a size 4 *Basic* design?

   (b) How many blocks are used in a size 10 *Basic* design?

   (c) If Jason made a size 50 *Basic* design, how many blocks would he need?

   (d) Explain how you can work out the number of blocks needed when you
        know the size number.

   (e) Choose one of the expressions below to complete the sentence:

        If the size number of a Basic design is *n* then the number of blocks in it is ...

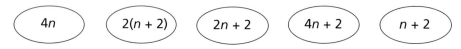

        $4n$ $2(n + 2)$ $2n + 2$ $4n + 2$ $n + 2$

2 This is Jason's *Petal* design.
   On the left is a size 2; on the right is a size 4.

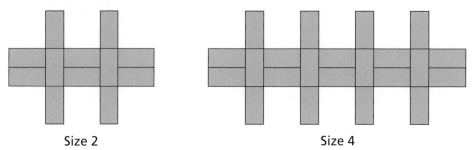

        Size 2   Size 4

   Which of these expressions tells you the number of blocks in a size *n* Petal design?

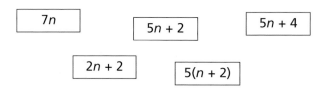

        $7n$ $5n + 2$ $5n + 4$

              $2n + 2$ $5(n + 2)$

Section D

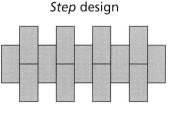

Step design

Size 4

1 In Jason's *Step* block paving design
   there are $3n + 1$ blocks in a size n design.

(a) Use the expression $3n + 1$ to
    check that in the size 4 *Step* design
    there are 13 blocks.

(b) Copy and complete this table for the *Step* paving design.

Size of *Step* design (n)	1	2	3	4	5
Number of blocks (b)				13	

(c) On squared paper draw axes going from 0 to 6 across and from 0 to 24 up.
    Label the across axis *Size of Step design (n)* and the up axis *Number of blocks (b)*.

(d) Plot points for the table above on your graph and join them with a straight line.

(e) Extend your line to find out how many blocks there are in a size 6 *Step* design.

2 This is one of Jason's *Cross* designs.
   It is a size 4, and uses 14 rectangular blocks.

Cross design

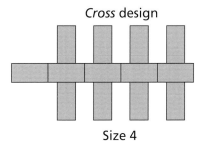

Size 4

(a) How many blocks would there be
    in a size 2 *Cross* design?

(b) Copy and complete this table
    for *Cross* paving designs.

Size of *Cross* design (n)	2	3	4	5
Number of blocks (b)			14	

(c) Which of these expressions is correct for the number of blocks (b)
    in a size n *Cross* design?

$4n - 2$		$3n + 1$		$3n + 2$		$4n - 1$

(d) On squared paper, draw axes going from 0 to 6 across and from 0 to 20 up.
    Label the across axis *Size of Cross design (n)* and the up axis *Number of blocks (b)*.

(e) Plot points for the table above on your graph and join them.

(f) Extend your line to find out how many blocks there are in a size 6 *Cross* design.

Section E

1 Lisa builds fences. She fixes together wooden posts to make the fences.
A size *n* Tribar fence has
$4n + 1$ posts in it.

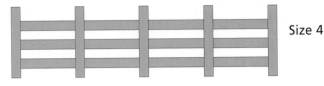

Size 4

(a) Check that the expression is correct for the size 4 fence above.

(b) One size of *Tribar* fence has 61 posts in it.
Copy and complete this working to find
out what size *Tribar* fence it is.

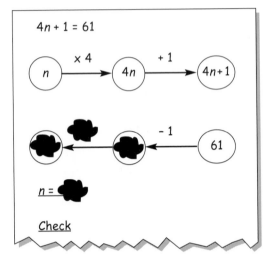

2 A size *n* Propped fence is made
from $4n + 3$ posts.

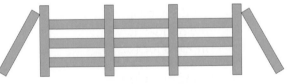

One Propped fence is made from 95 posts.
Work out what size it is.
Show all your working and check your answer.

Size 3

3 Below are three sizes of Lisa's *Hurdle* fence.

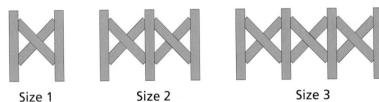

Size 1   Size 2   Size 3

(a) Which of these expressions tells you
the number of posts in a size *n* Hurdle fence?

(b) Lisa makes a *Hurdle* fence with 82 posts.
Work out what size it is.

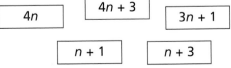

30

Mixed questions 1

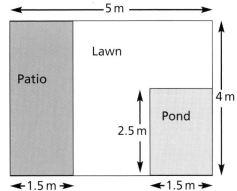

1 This diagram shows a garden. Find the area of

  (a) the whole garden

  (b) the patio

  (c) the pond

  (d) the lawn

2 If edging was put around the perimeter of the lawn, how much would be needed?

3 This graph shows the midday temperature in Paris and Moscow for the first week of January.

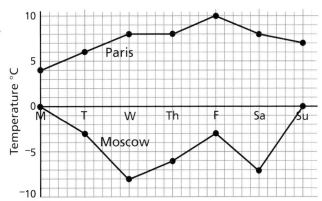

  (a) What was the temperature at midday in Moscow on Friday?

  (b) On what day was the difference in temperature between the two cities greatest?

  (c) Find the mean and range of the midday temperatures in Paris for this week.

4 Copy and complete each of these.

  (a)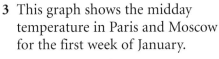
  (b)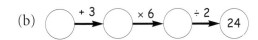

5 Draw an arrow diagram for each of these equations.
Reverse each diagram to solve the equation. Check your answers work.

  (a) $3x + 2 = 26$ (b) $4w - 5 = 23$ (c) $\frac{k}{5} + 3 = 6$

6 This diagram shows part of a shape drawn on a coordinate grid.
The whole shape has one line of symmetry, the *y*-axis.

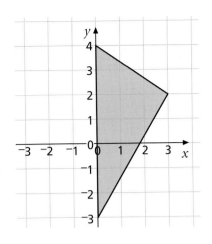

(a) Copy the diagram.
Add another point and complete the shape.

(b) Write down the coordinates of the new point.

(c) What is the name of this type of shape?

(d) What is the order of rotation symmetry of the complete shape?

7 Here are three of Barry's necklaces.

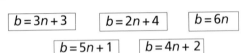

Size 1
Size 2
Size 3

(a) How many beads would be needed to make a size 4 necklace?

(b) Copy and complete this table.

Size of bracelet (n)	1	2	3	4	5	10
Number of bars (b)	6	11				

(c) Which of these expressions is correct for the number of beads (*b*) in a size *n* necklace?

$b = 3n + 3$ $b = 2n + 4$ $b = 6n$
$b = 5n + 1$ $b = 4n + 2$

(d) What size necklace will use 41 beads?
Show all your working.

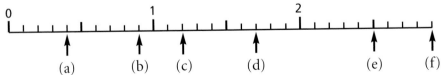

8 (a) You come out of St Mary's hospital and turn left into Praed Street. Which compass direction is that?

(b) From Talbot Square you go north-east into Sussex Gardens. You take the second left and first right. Where are you?

(c) Give directions to get from Paddington Tube station (marked ⊖) to Radnor Place.

9 What number does each of these arrows point to?

0   1   2

(a)   (b)   (c)   (d)   (e)   (f)

10 Put each of these sets in order, smallest first.

(a) 5.5, 5.25, 5.03, 5.05, 5.52

(b) 1, 0.08, 0.25, 0.05, 0.5, 0.85

(c) 250 g, 0.75 kg, $\frac{1}{2}$ kg, 1.5 kg

(d) 5 km, 500 m, 0.05 km, 900 cm

10 Experiments

Sections B and C

1 Class 10J carried out a memory experiment using a list of words.
The numbers of words that the students remembered were

8, 5, 5, 7, 8, 10, 9, 10, 9, 8, 9, 10, 9, 7, 8, 10, 7, 7, 8, 9, 10, 9, 9, 7, 8

(a) How many people in the class remembered 10 words?

(b) What was the frequency of remembering 6 words?

(c) Which number of words was remembered with a frequency of 6?

(d) Draw a dot plot of the number of words remembered.

(e) What is the modal number of words remembered?

2 The same class were given some pictures to remember.
Here are the numbers of pictures the students in the class remembered.

7, 6, 3, 7, 7, 10, 10, 10, 9, 8, 9, 10, 10, 9, 9, 10, 8, 8, 7, 10, 10, 9, 10, 9, 10

(a) Make a frequency table of these results.

(b) Draw a bar chart of the results.

(c) What is the frequency of remembering 8 pictures?

(d) What is the modal number of pictures remembered?

3 Look at the number of pictures remembered in question 2.

(a) Write down the numbers remembered in order.

(b) Use your list to find the median of the number of pictures remembered.

(c) What is the range of the number remembered?

4 The same class were given some two-digit numbers to remember.
This is how many numbers each person remembered.

5, 6, 7, 5, 7, 8, 8, 9, 10, 6, 6, 9, 8, 9, 8, 7, 4, 10, 8, 7, 9, 6, 10, 4, 7

(a) Make a frequency table for these results.

(b) Use your frequency table to draw a bar chart.

(c) Find the median and range of how many two-digit numbers
each person remembered.

5 Put the class results from questions 1, 2 and 4 into a table like this.

Type	Median	Range
Words		
Pictures		
Numbers		

Use your table to say whether these statements are true or false.

(a) *Pictures were remembered best of all as they had the highest median.*

(b) *Words had the smallest range, so these were the hardest to remember.*

(c) *Numbers were the hardest to remember as the median was the smallest.*

6 In another experiment the students in a class were asked to guess the age of their teacher. The answers they gave were

Girls: 30, 32, 33, 30, 28, 31, 36, 29, 32, 32, 30, 29, 30, 31, 29

Boys: 32, 29, 39, 30, 32, 32, 26, 33, 35, 31, 30, 32, 31, 33, 30, 35, 37

(a) Draw a dot plot showing the guesses of the girls.

(b) Draw another dot plot, using the same scale, to show the boys' guesses.

(c) Find the median and range of the girls' guesses.

(d) Find the median and range of the boys' guesses.

(e) Which group, on average, thought the teacher was older, boys or girls?

(f) Whose guesses varied the most, boys' or girls'?

7 In the memory experiment for question 1 the number of words remembered was listed separately for boys and girls.

Boys: 5, 5, 9, 8, 9, 9, 7, 10, 8, 8, 9, 9, 9, 7, 8

Girls: 8, 7, 8, 10, 10, 9, 10, 7, 7, 10

(a) Find the mean number of words remembered by the boys.

(b) Find the mean number of words remembered by the girls.

(c) Who remembered the most words, on average, boys or girls?

8 A group of students carried out an experiment where they tested their grip strength.
First they tested their grip using the hand they write with.
They then tested with their 'non-writing' hand.

The results were

Grip strength with writing hand (kg):   29 35 34 39 41 43 37 35 30 36

Grip strength with non-writing hand (kg):   28 26 48 27 32 36 24 28 31 27

(a) Make a copy of the scale on the right.
Use it to draw a dot plot for the students'
grip strength with their writing hand.

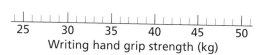

(b) Find the median of the grip strengths with their writing hand.

(c) Draw a dot plot for the students' grip strength with their
non-writing hand.
Use the same size scale as before.

(d) Find the median of the students' grip strengths with their
non-writing hand.

(e) Which hand, on average, had the strongest grip,
the writing hand or the non-writing hand?

9 A class were given ten different angles to estimate.
They recorded the number of angles each person estimated correctly within 5°.

7, 6, 4, 8, 5, 6, 3, 7, 6, 5, 6, 6, 4, 6, 5, 8, 2, 8, 6, 7

(a) Find the mean number of angles a person estimated correctly.

After using a computer program to practise, they were given
another ten angles to estimate.
These were the results.

8, 7, 6, 6, 8, 9, 10, 6, 7, 9, 8, 7, 7, 8, 6, 8, 10, 9, 8, 7

(b) Find the mean number of angles a person estimated correctly after practice.

(c) Do you think that the practice helped the students to estimate angles?
Give your reason.

11 Estimating and scales

Section A

1 The length of this drinking straw is 5 cm.

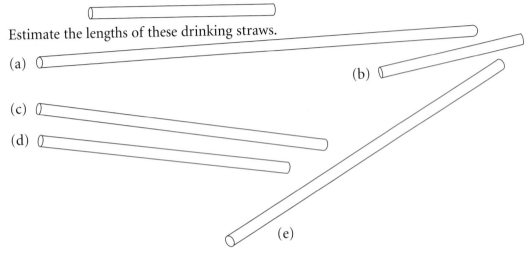

Estimate the lengths of these drinking straws.

(a)

(b)

(c)

(d)

(e)

2 The diagram shows the plan of a tennis court.

The tennis court is 10 m wide. Estimate its length.

10 m

3 The height of the hippopotamus in this picture is 1.5 m.

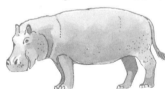

Estimate the height of
(a) the elephant
(b) the hyena
(c) the giraffe

Section B

1 These diagrams show a set of scales measuring in grams.
   What weight does the pointer show?

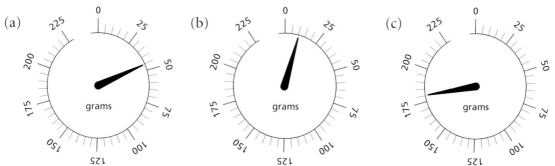

(a)   (b)   (c)

2 This is the dial from some bathroom scales.
   It measures weight in kilograms.

   (a) What does each small division represent?

   (b) The 'pointer' is the vertical line.
        What weight is shown on the scales?

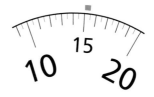

(i)   (ii)   (iii)

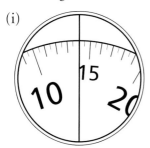   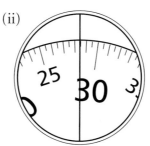   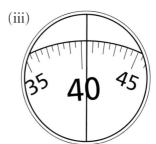

3 These diagrams show the scale on a spring balance
   used in science lessons.

   (a) What does each small division represent?

   (b) What weight is shown in each case?

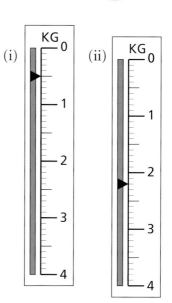

(i)   (ii)

12 *Into the crowd*

Sections A, B and C

1

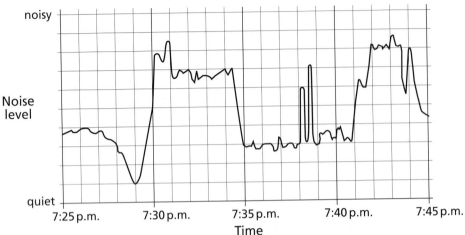

The noise at the pantomime

The graph shows the noise levels at the start of a pantomime.

(a) The audience went very quiet as the lights were dimmed before the start.
    When was this?

(b) The orchestra started the pantomime by playing an overture
    just before the narrator came on stage.
    For about how long were the orchestra playing?

(c) The narrator talked to the audience for about 3 minutes.
    Then she encouraged them to practise shouting at the 'Baddie'.
    When did they shout at the 'Baddie'?

(d) The narrator talked a little longer and then she sang a song. It was quite loud!
    When did she start singing the song?

(e) At the end of the song the audience applauded loudly.
    When did the audience applaud?

2 Harestead village needed a bypass. The residents found it very difficult crossing the road and it was very noisy.

The Council organised traffic surveys. They counted the vehicles that went through the village every hour. They produced these graphs for their report.

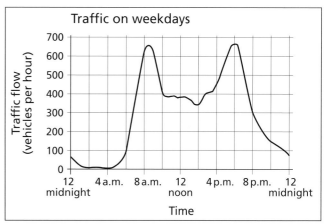

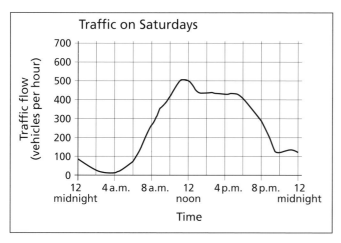

(a) About how many vehicles were passing per hour at 8 a.m. on a weekday?

(b) Describe the traffic flow through the village on weekdays.

(c) When was the village busiest on a Saturday?

(d) The data shows the traffic flow, in vehicles per hour, for Sundays.

Time	midnight	1 a.m.	2 a.m.	3 a.m.	4 a.m.	5 a.m.	6 a.m.	7 a.m.	8 a.m.	9 a.m.	10 a.m.	11 a.m.
Traffic flow	120	70	30	20	10	20	40	80	130	230	360	460
Time	noon	1 p.m.	2 p.m.	3 p.m.	4 p.m.	5 p.m.	6 p.m.	7 p.m.	8 p.m.	9 p.m.	10 p.m.	11 p.m.
Traffic flow	510	440	400	450	470	460	430	380	310	240	180	110

(i) Use the data to draw a similar graph for Sundays.

(ii) How does Sunday traffic differ from Saturday traffic?

13 *Imperial measures*

Sections A and B

1 A garden measures 30 feet by 75 feet.
What are these measurements in metres, approximately?

2 Roughly how many feet are there in

   (a) 2 m   (b) 50 m   (c) 200 m   (d) 1 km

3 For each of these pairs, say which distance is longer.

   (a) 9 kilometres   5 miles   (b) 30 kilometres   20 miles

   (c) 100 km   100 miles   (d) 16 km   12 miles

4 Change these speed limits in miles per hour to kilometres per hour.

   (a)   (b)   (c)

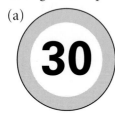 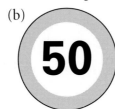 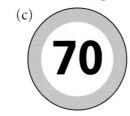

5 (a) Round these distances from Manchester to the nearest 10 miles.

        Birmingham   81 miles   Derby   58 miles

        Leeds   39 miles   Brighton   247 miles

   (b) Roughly what are these distances in kilometres?

6 What are these weights roughly in kilograms?

   (a) 1 lb   (b) 4 lb   (c) 20 lb   (d) 100 lb

7 What are these weights roughly in pounds (lb)?

   (a) 2 kg   (b) 10 kg   (c) 0.25 kg   (d) 1.5 kg

8 Which is the heavier in each of these pairs?

   (a) 5 lb   5 kg   (b) 12 lb   7 kg

   (c) 80 lb   30 kg   (d) $\frac{1}{2}$ lb   300 g

9 Here is a shopping list from Great Grannie's Restaurant.
Write the list out using kilograms instead of pounds.

Potatoes	30 lb
Sprouts	10 lb
Sugar	8 lb
Salt	3 lb

14 Rounding and multiplying

Section A

1 Which whole number is nearest to each of these?

(a) 28.6   (b) 7.3   (c) 5.2   (d) 18.5   (e) 9.8

2 Round these to the nearest centimetre.

(a) 3.7 cm   (b) 9.4 cm   (c) 8.2 cm   (d) 19.8 cm   (e) 20.9 cm

3 Round these to the nearest kilometre.

(a) 124.3 km   (b) 98.9 km   (c) 342.2 km   (d) 983.7 km   (e) 127.9 km

4 Which whole number is closest to each of these?

(a) 23.82   (b) 29.88   (c) 34.25   (d) 48.65   (e) 50.36

5 Round these to the nearest kilogram.

(a) 6.54 kg   (b) 3.09 kg   (c) 13.43 kg   (d) 8.84 kg   (e) 10.75 kg

6 Round these to the nearest whole number.

(a) 7.8   (b) 42.56   (c) 15.48   (d) 29.69   (e) 0.76

Section B

1 Round these numbers to one decimal place.

(a) 15.63   (b) 7.893   (c) 0.95   (d) 10.745   (e) 1.405

2

A	E	C	K	R	H	S	B	D
4.0	4.1	4.2	4.3	4.4	4.5	4.6	4.7	4.8

Round each decimal below to one decimal place and find a letter for each one. Rearrange each set of letters to spell an item of food.

(a) 4.0937, 4.03, 4.1875, 4.3453

(b) 4.477, 4.134, 4.059, 4.567, 4.235, 4.088

(c) 4.392, 4.778, 4.7341, 4.079, 4.032

3 Round these numbers to one decimal place.

(a) 35.7326   (b) 0.471 34   (c) 7.381 52   (d) 2.134 15

Section C

1. Round these numbers to two decimal places.
   - (a) 4.9453
   - (b) 9.0673
   - (c) 24.089
   - (d) 1.338
   - (e) 47.938
   - (f) 14.634
   - (g) 28.149
   - (h) 4.299

2. Round these amounts to the nearest penny.
   - (a) £5.9456
   - (b) £20.949
   - (c) £6.079
   - (d) £32.499

3. Round these numbers to three decimal places.
   - (a) 1.234 56
   - (b) 3.567 85
   - (c) 0.885 43
   - (d) 17.8999

4. The numbers in the rectangles are written to three decimal places in the loops. Find three matching pairs.

Section D

1. Calculate the following.
   - (a) 2.57 × 10
   - (b) 0.738 × 100
   - (c) 58.678 × 1000
   - (d) 9.842 × 100
   - (e) 2.38 × 10 000
   - (f) 1.7865 × 100

2. Calculate the following.
   - (a) 4.8 ÷ 10
   - (b) 369.32 ÷ 100
   - (c) 68.47 ÷ 1000
   - (d) 9.7 ÷ 100
   - (e) 140 ÷ 10 000
   - (f) 0.9 ÷ 10

3. Calculate the following.
   - (a) 76.3 × 100
   - (b) 0.7861 × 1000
   - (c) 0.987 ÷ 100
   - (d) 0.765 × 1000
   - (e) 10.769 ÷ 100
   - (f) 76.93 × 10 000

4. Find the missing number in each calculation.
   - (a) ■ × 10 = 75.3
   - (b) 7321 ÷ ■ = 73.21
   - (c) 1000 × ■ = 45.63
   - (d) 17.3 × ■ = 17 300

5. Copy and complete these chains.

   (a)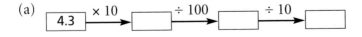

   (b)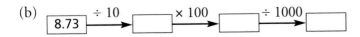

Section E

1 Change these lengths to kilometres.
   (a) 73 600 m   (b) 9 631 200 m   (c) 480 m   (d) 4320 m   (e) 63 m

2 Change these lengths to centimetres.
   (a) 63 mm   (b) 0.12 m   (c) 8 mm   (d) 83.6 m   (e) 0.089 m

3 Change these lengths to metres.
   (a) 68 cm   (b) 7.93 km   (c) 8456 mm   (d) 0.095 km   (e) 6400 cm

4 Change these weights to kilograms.
   (a) 6392 g   (b) 3920 g   (c) 750 g   (d) 800 g   (e) 50 g

5 Change these volumes to millilitres.
   (a) 8 litres   (b) 4.926 litres   (c) 18.3 litres   (d) 9.05 litres   (e) 0.09 litres

6 Put these weights in order, smallest first.
   560 g, 0.763 kg, 77 g, 1.055 kg, 0.553 kg, 0.05 kg, 0.5 kg

7 I walk 3.1 km to work and then 1500 m to the shops.
   How far have I walked altogether in kilometres?

8 I buy 6 packets of dried fruit to make a Christmas cake.
   Each packet holds 200 grams.
   How many kilograms of dried fruit have I bought?

9 Human hair grows at an average rate of 7.5 mm a month.
   How much, in **centimetres**, would hair grow on average
   (a) in 10 months   (b) in 100 months (about 8 years)

10 Susan has a 2 kg bag of flour.
    How much does she have left if she uses
    (a) 250 g   (b) 850 g   (c) 0.75 kg   (d) 1 kg 250 g

15 *Evaluating expressions*

Section A

⊠ 1 Without using a calculator, evaluate each of these.

    (a) $6 + 4 \times 2$ (b) $(7 + 4) \times 2$ (c) $5 \times (8 - 3)$ (d) $6 \times 3 - 7$

    (e) $10 - (4 + 2)$ (f) $8 - (5 - 3)$ (g) $\dfrac{12 + 9}{3}$ (h) $10 + \dfrac{6}{3}$

    (i) $\dfrac{15}{3} + 2$ (j) $\dfrac{10 - 4}{2}$ (k) $8 - \dfrac{9}{3}$ (l) $\dfrac{16}{4} - 3$

2 Find the missing number in each of these calculations.

    (a) $3 \times 5 - \blacksquare = 9$ (b) $\dfrac{\blacksquare + 11}{4} = 5$ (c) $(4 + \blacksquare) \times 5 = 35$

3 Evaluate each of these. (Use a calculator when you need to.)

    (a) $(2 + 4.5) \times 3$ (b) $\dfrac{(7 + 4)}{2}$ (c) $\dfrac{8}{5} + 2$ (d) $2.5 \times (6 - 3)$

    (e) $8 - \dfrac{4.5}{1.5}$ (f) $3.2 \times 4 + 7$ (g) $\dfrac{13 + 8}{3.5}$ (h) $6.8 - \dfrac{14}{7}$

Section B

1 What is the value of each expression when $p = 8$?

    (a) $p + 6$ (b) $\dfrac{p}{2}$ (c) $4p$ (d) $p - 5$

    (e) $\dfrac{p + 4}{4}$ (f) $6(p + 1)$ (g) $\dfrac{p}{2} - 2$ (h) $4(p - 3)$

2 What is the value of each expression when $x = 18$?

    (a) $\dfrac{x - 9}{3}$ (b) $3(x - 10)$ (c) $3x$ (d) $\dfrac{x}{3} - 2$

3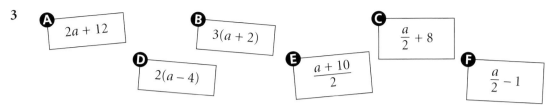

(a) Work out the value of each expression when $a = 20$.

(b) Which expression has the value 0 when $a = 4$?

(c) Which expression has the smallest value when $a = 10$?

(d) Which two expressions have the same value when $a = 6$?

4 Each expression in the diagram stands for the length of a side in metres.

(a) Work out the length of each side when $d = 5$.

(b) Sketch the rectangle and label the lengths of the sides.

(c) What is the perimeter of your rectangle?

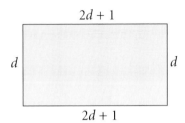

Section C

1 *Digby's Drawers* make cabinets.
He use this rule to work out the price in pounds.

   price = 20 × number of drawers + 35

Use their rule to work out the price of a cabinet with

(a) 2 drawers (b) 6 drawers

2 *Pat's Pizzas* use this rule to work out the cost of their pizzas in pence.

   cost = 25 × number of toppings + 350

Work out the cost of a pizza with

(a) 4 toppings (b) 7 toppings

3 Kirk uses this rule to estimate how long a walk will take him in minutes.

   time = 12 × number of kilometres + 30

Roughly how long will Kirk take to walk 8 kilometres?

4 *Youth Activities* use this rule to find the number of instructors they need for a group of students.

   number of instructors = $\dfrac{\text{number of students}}{4}$ + 1

How many instructors do they need for 20 students?

5 You can use this rule to calculate the distance travelled on a journey.

distance = speed x time

(a) Kim walked at a speed of 4 km/h.
She walked for 3 hours.
Work out the distance she walked.

(b) How far can Dominic cycle if his speed is 12 km/h and
he cycles for 3 hours ?

Section D

1 At *Cycle Hire* they charge £7 an hour.
You can use this formula to work out how much to pay.

$$C = 7H$$

C is the total cost and *H* is the number of hours.

What is the total cost if you hire a bike for

(a) 3 hours   (b) 6 hours

2 A fabric shop uses the following rule to work out
what length of fabric is needed for some curtains.

$$L = 4H + 60$$

L is the length of fabric in cm and *H* is the height of the window in cm.

What length would you need if the height of the window is

(a) 255 cm   (b) 358 cm

3 Sam uses this rule to work out the height of a roof.

$$H = \frac{R}{2} + 25$$

H is the height of the roof in cm and
R is the length of the rafter in cm.
Calculate the height of the roof if the rafter is

(a) 350 cm   (b) 486 cm

Rafter

Height

4 Alex uses this rule to work out the
total height of a stack of chairs

$$H = 11N + 84$$

H is the total height in cm and
N is the number of chairs in the stack.

What is the total height of

(a) 2 chairs   (b) 7 chairs

16 Shopping

Use a pie chart scale for this page.

Section B

1 This pie chart shows the costs of keeping a car on the road.

   (a) What percentage of the cost of keeping a car is insurance?

   (b) What percentage of the cost of keeping a car is breakdown cover?

   (c) What percentage of the cost of keeping a car is tax?

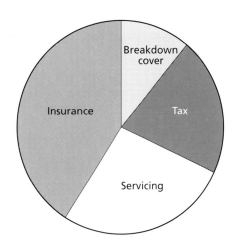

2 This pie chart shows how the maths department spent its budget last year.

   (a) What did the maths department spend the most money on?

   (b) What percentage of the budget was spent on consumables?

   (c) What percentage was spent on equipment?

   (d) If the department budget was £8000, roughly how much was spent on exercise books and paper?

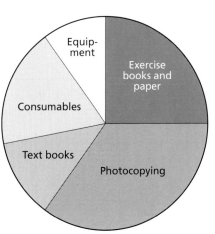

Section C

1 This information shows the contents of 3 different types of chocolate.

	White chocolate	Milk chocolate	Plain chocolate
Protein	6%	8%	4%
Carbohydrate	57%	57%	65%
Fat	34%	30%	30%
Water	3%	5%	1%

Draw a pie chart with a radius of 4 cm for each type of chocolate.

2 This information shows how a children's charity spends its income.

UK work	Third world	Fund-raising	Administration	Other
27%	42%	16%	11%	4%

Draw a pie chart with a radius of 4 cm to show how the charity spends its money.

Section D

1 This chart shows the percentage of households which owned
various types of electrical goods in the UK in 1996 and 1999.

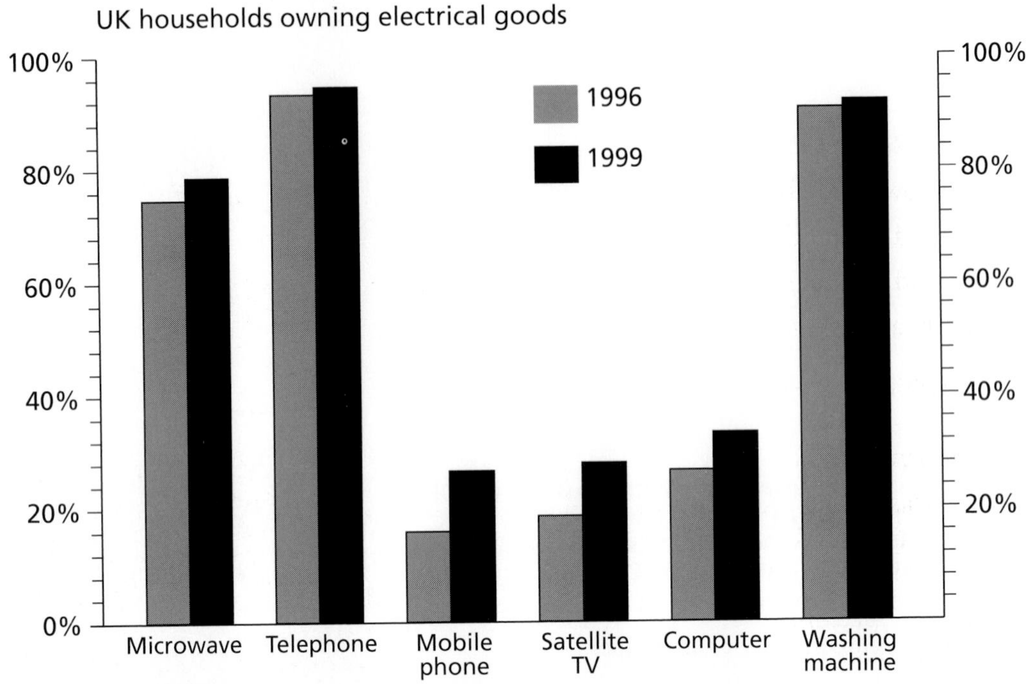

UK households owning electrical goods

(a) Which item saw the biggest increase between 1996 and 1999?

(b) What percentage of households had satellite television in 1996?

(c) What percentage of households had a computer in 1999?

(d) Which items saw the smallest change between 1996 and 1999?

2 (a) These figures show the spending of two different charities.
Use this information to draw a bar chart like the one above.
comparing the spending of the two charities.

	Emergencies	Health	Education	Fund-raising	Administration
Charity A	16%	29%	31%	22%	2%
Charity B	25%	38%	15%	14%	8%

(b) Describe the main differences between the spending of the two charities.

Mixed questions 2

1 Some pupils counted how many plastic cubes they could pick up in one hand.
They each tried it with their right hand, and then their left hand.

The results were   Right hand   Left hand
                           4 8 6 6 7 6 9 5   6 5 4 6 6 6 6 5
                           7 7 6 5 7 7 8   7 7 6 5 7 5 7

(a) Draw a bar chart for the number of cubes picked up in the right hand.

(b) Draw a bar chart for the number of cubes picked up in the left hand.

(c) Find the modal number of cubes picked up for each hand.

(d) Find the median and range of the number of cubes for each hand.

(e) On average, which hand could pick up more cubes?

2 You will need a pie chart scale for this question.
This pie chart shows the reasons people
complained about adverts on television.

(a) Measure the percentage of people who
complained because they thought an advert
was misleading.

(b) What percentage complained about
harmful adverts?

(c) In 1998 there were around 8000 complaints
about adverts.
Roughly how many were because people
thought they were offensive?

3 A cycling club uses this rule to find out how long a cycling tour takes.

$T = \dfrac{d}{20} + 2$ (T is the time in hours, d is the distance in kilometres)

(a) Use this rule to find the time taken for tours of
(i) 80 km   (ii) 100 km   (iii) 50 km

(b) Copy and complete this table.

Distance (d) km	40	80	120	160	200
Time (T) hours					

(c) What distance could be travelled on a nine hour tour?

4 Round these quantities.

(a) 15.37 kg to the nearest kilogram

(b) 12.2375 to two decimal places

(c) 0.876 65 to three decimal places

5 What is the weight of each of these animals?

(a) (b) (c)

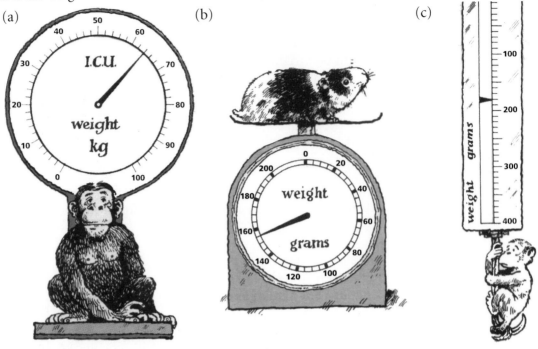

6 Food hall customers: Wednesday

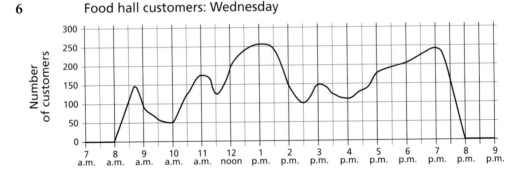

A Food Hall contains cafés, fast food shops and sandwich bars.
The graph shows the number of customers on a Wednesday.

(a) At what times does the Food Hall open and close?

(b) Why do you think the Food Hall was fairly busy between 8 a.m. and 9 a.m.?

(c) When are the staff in the Food Hall least busy?

(d) How many customers were in the Hall when it was busiest?

(e) Morning coffee is popular. When do you think it was served?

(f) The Food Hall was busy before people went home in the evening.
When was this?

(g) When did the number of customers in the Hall increase the most quickly?

(h) How many customers were in the Hall at 3 p.m.?

18 *Drawing and using graphs*

Section A

1 Amanda is collecting liquid in a chemistry experiment.
When she starts timing she has 2 ml of liquid in the cylinder.
Each minute she collects 3 more millilitres of liquid.

(a) How much liquid will she
have after 2 minutes?

(b) Copy and complete this table.

Time in minutes	0	1	2	3	4	5
Liquid collected in ml	2	5				

(c) On graph paper draw and label axes like these.
Then plot the points from your table.

(d) Join the points you have plotted.
Extend the line they make.

(e) Use your graph to say how much liquid
she will have after 7 minutes.

(f) How many minutes will it take until she
has 20 ml of liquid?

(g) The cylinder she is collecting the liquid in
can only hold 25 ml.
How long will it be until the cylinder is full?

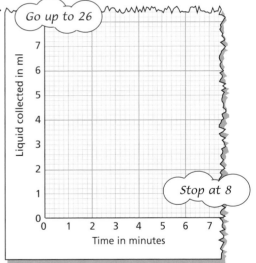

51

2 A pond must be emptied before cleaning.
It contains 1200 litres of water.
60 litres of water are pumped out
of the pond every minute.

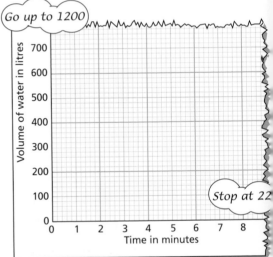

(a) How much water is left
   in the pond after 2 minutes?

(b) How much water is left
   in the pond after 10 minutes?

(c) Copy and complete this table.

Time in minutes	0	2	4	6	8	10
Volume of water in litres	1200					

(d) On graph paper, draw and label axes like these.
   Plot the points from your table.

(e) Join the points you have plotted with a line
   and extend it.

(f) Use your graph to say how much water
   is left in the pond after 15 minutes.

(g) After roughly how many minutes will
   there be 500 litres left in the pond?

(h) How long will it take to empty the pond?

(i) At what time will the pond be empty,
   if the pumping started at 10:45 a.m.?

Section B

1 *Highrise Construction* uses this rule to work out how many days holiday its employees are allowed each year.

number of days holiday = (number of years employed ÷ 2) + 18

(a) Use the rule to check that someone employed for 4 years will get 20 days holiday.

(b) Work out the number of days holiday for someone who has been employed for

    (i) 2 years   (ii) 10 years

(c) Copy and complete this table.

Number of years employed	0	2	4	6	8	10
Number of days holiday	18		20			

(d) Plot the points from your table using axes like these.

(e) Join the points with a line and extend it.

(f) How many days holiday would you get if you had been employed for 12 years?

(g) For how long have you been employed if you get 26 days holiday?

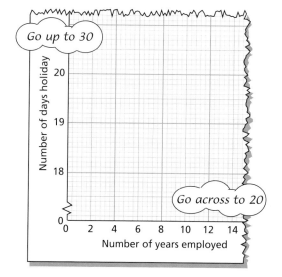

(h) Copy and complete these holiday records.

    (i)

> **George Jones**
>
> _____ years
>
> **27 days holiday**

    (ii)

> **Kay Harvey**
>
> 11 years
>
> __ days holiday

    (iii)

> **Abdul Hussain**
>
> _____ years
>
> $19\frac{1}{2}$ days holiday

    (iv)

> **Jack North**
>
> 5 years
>
> __ days holiday

2 *Speedyfix* uses this formula to calculate the charge for a repair job.

$c = 35 + 10h$

c is the charge in pounds,
h is the number of hours the job takes.

For example, if a job takes 2 hours, $c = 35 + 10 \times 2 = 55$
so the charge for the job would be £55.

(a) Copy and complete this table showing the charge
    for different lengths of job.

Hours taken (h)	1	2	3	4	5
Charge in £ (c)		55			

(b) Draw a graph to show the values in your table.
    Draw and label your axes like this.
    Plot the points from your table
    and join them with a line.
    Extend your line.

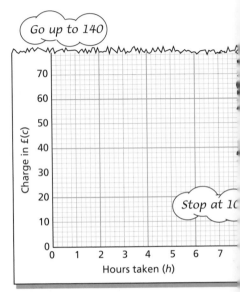

(c) Use your graph to complete these bills.

(i)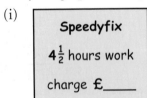

(ii)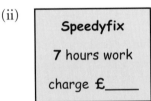

(iii)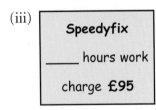

(iv)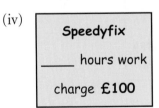

(v)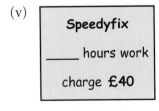

(vi)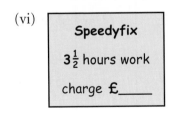

19 *Written adding and subtracting*

Sections A and B

1 (a) Use the digits 3, 7, 6, 2 in this addition to make the largest total possible.

(b) Make the largest **difference** that you can in this subtraction using the digits 3, 7, 6, 2.

2 In a magic square the numbers in each **row**, **column** and **diagonal** add to the same total. Copy and complete these magic squares.

(a)
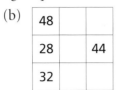

11		21
	15	
		19

(b)

48		
28		44
32		

(c)
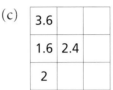

3.6		
1.6	2.4	
2		

3 Work out

(a) 36 + 48
(b) 156 + 385
(c) 509 + 294
(d) 83 − 48
(e) 126 − 72
(f) 431 − 275

4 (a) Here are four guesses of how many sweets there are in this jar.

| 916 | | 783 | | 870 | | 796 |

The actual number of sweets is 834. Which guess is closest?

(b) These guesses are for the weight of a cake.

| 3.25 kg | | 4 kg | | 3.9 kg | | 3.4 kg |

Its actual weight is 3.67 kg. Which guess is closest?

5 Work out in your head

(a) 3 + 2.6
(b) 4.8 − 2.3
(c) 8.1 + 3
(d) 4.6 − 2
(e) 4.6 + 5.4
(f) 5 − 0.4
(g) 1.5 + 1.25
(h) 2 − 1.2

6 Work out (a) 4.5 litres + 1.25 litres (b) 3 kg − 1.45 kg (c) 3.7 m − 1.24 m

7 Whose shopping is heavier, Jo's or David's? By how much?

Jo

0.5 kg
2.5 kg
3.15 kg
1.76 kg

David

0.35 kg
2.65 kg
1.08 kg
3.25 kg

Section C

1 Find the perimeters of these shapes.

(a)

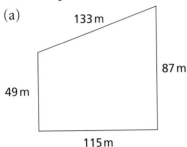

(b)

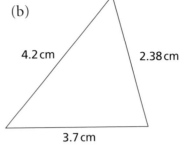

(c)

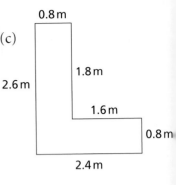

2 A triangular field has perimeter 560 m.
One side has length 213 m and another has length 148 m. How long is the third side?

3 (a) Angles on a straight line add up to 180°.
Work out the missing angle.

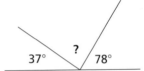

(b) Angles at a point add up to 360°.
Calculate the missing angle.

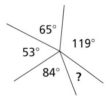

4 (a) The sum of the angles of a triangle is 180°.
Find the missing angle.

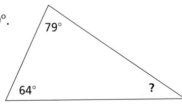

(b) The sum of the angles of a quadrilateral is 360°.
Find the missing angle.

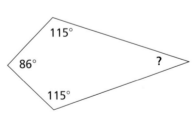

5 Here are the lengths of some leaves.

3.6 cm   2.7 cm   3.9 cm   4.25 cm   2.85 cm

The range is the difference between the longest and shortest lengths.
Find the range of the lengths.

20 *Frequency*

Sections A and B

1 This data is the number of visitors to an exhibition each day in February.

24, 36, 41, 17, 25, 36, 22,
13, 49, 34, 16, 23, 19, 25,
32, 47, 37, 15, 27, 29, 31,
28, 47, 39, 16, 21, 30, 19

(a) Copy and complete this stem-and-leaf table for the data.
(The first two pieces of data have already been put in the table.)

```
0 |
1 |
2 | 4
3 | 6
4 |
5 |
```

Stem = 10 people

(b) Copy out your table putting the 'leaves' in order.

(c) Find the range of the number of people.

(d) What was the median number of people?

2 This table shows the number of visitors in March.

```
1 | 1 2 2 2 4 5 6 7 7 8 8
2 | 2 3 4 4 5 6 6 6 7 9 9
3 | 0 2 2 3 6 7
4 | 1 4
5 | 2
```

Stem = 10 people

Find the median and range of the number of visitors in March.

Section C

1 James is comparing two different types of tomato plant.
He counts the number of tomatoes he picks from each plant.
This table shows the results for the two types of plant.

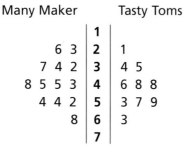

Many Maker   Tasty Toms

	1	
6 3	2	1
7 4 2	3	4 5
8 5 5 3	4	6 8 8
4 4 2	5	3 7 9
8	6	3
	7	

Stem = 10 tomatoes

(a) Find the median and range of the number of tomatoes from Many Maker plants.

(b) Find the median and range of the number of tomatoes from Tasty Toms plants.

(c) Decide whether each of these statements is true or false.

    A   The median for Many Maker is higher so they produce more tomatoes on average.

    B   The highest number of tomatoes was picked from a Tasty Toms plant so they produce more on average.

    C   The range for Tasty Toms plants was lower so Tasty Toms produce more tomatoes on average.

2 Mario counts the number of aubergines on two types of plant.
This table shows the number of aubergines on the two types of plant.

Black Beauty   Long Purple

9 6	0	7 8 8
8 8 3 1 1	1	0 1 2 6 7
9 9 7 4	2	1 3 6 8
4 0	3	

Stem = 10 aubergines

(a) Find the median and range for each type of aubergine plant.

(b) Which type of plant gave Mario more aubergines on average?

Section D

1 Rashid runs a video hire shop called *Visual Videos*. The graph shows the ages of people who went to *Visual Videos* on a Tuesday.

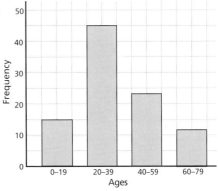

(a) How many people aged 40–59 went to *Visual Videos* on Tuesday?

(b) How many people aged 0–19 went to *Visual Videos* on Tuesday?

(c) What was the modal age group for Tuesday?

(d) How many people in total visited *Visual Videos* on Tuesday?

2 This table shows the ages of people who visited *Visual Videos* on the previous Saturday.

Age group	Frequency
0–19	25
20–39	44
40–59	51
60–79	10

(a) Copy and complete this bar chart to show the data in the table.

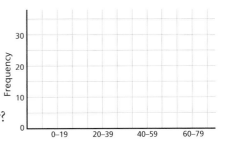

(b) (i) How many people in total visited *Visual Videos* on Saturday?

(ii) How many more people visited *Visual Videos* on Saturday than on Tuesday?

(iii) Why do you think the numbers of visitors are so different on the two days?

(c) How many people over the age of 19 went to *Visual Videos* on Saturday?

(d) Give two differences between the people who went to *Visual Videos* on Tuesday and those who went on Saturday.

21 Number links

Section A

1 Which numbers in this list are multiples of 7?

   42, 24, 14, 17, 18, 700, 71, 35

2 Which numbers in this list are multiples of 3?

   23, 12, 9, 15, 31, 300, 28, 24

3 Write down six different multiples of 6 which are less than 64.

4 Write down six different multiples of 9 which are less than 100.

5 Write down all the multiples of 8 which lie between 30 and 50.

6 Which numbers in the loop are

   (a) multiples of 7   (b) multiples of 2

   (c) multiples of 11   (d) multiples of 5

   (e) multiples of 4   (f) multiples of **both** 5 and 4

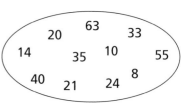

Section B

1 One number in this list is not a factor of 12. Which is it?

   2, 3, 4, 5, 6, 12

2 Two numbers in this list are not factors of 18. Which are they?

   2, 3, 4, 6, 8, 9

3 List all the factors of

   (a) 8   (b) 12   (c) 30   (d) 16   (e) 13

4 Which of these are common factors of 12 and 32?

   1, 2, 4, 5, 6, 8, 9, 10,

5 List the common factors of

   (a) 12 and 18   (b) 10 and 15   (c) 8 and 20

6 Show how you can fit the numbers 3, 4, 5 and 8 into this grid, one number in each box.

	is a factor of 24	is a factor of 20
is a factor of 40		
is a factor of 12		

Section C

1 Decide which of these statements are true and which are false.

  (a) 3 is a factor of 12.   (b) 9 is a multiple of 3.

  (c) 15 is a factor of 5.   (d) 5 is a factor of 25.

  (e) 18 is a multiple of 6.   (f) 24 is a multiple of 8.

  (g) 16 is a factor of 4.   (h) 12 is a multiple of 3.

2 Which word, either 'factor' or 'multiple', should go in each statement?

  (a) 4 is a .... of 24.   (b) 7 is a .... of 28.

  (c) 15 is a .... of 3.   (d) 9 is a .... of 27.

  (e) 9 is a .... of 3.   (f) 15 is a .... of 45.

Section D

1 List all the prime numbers between 20 and 30.

2 Four of the numbers in the loop are not prime. Which are they?

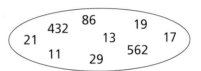

3 Which of the numbers below are prime?

   12, 7, 48, 29, 5, 23

4 List all the prime numbers between 40 and 50.

5 How can you tell that none of the numbers below are prime?

   4, 56, 24, 128, 16, 34

Section E

1 Which of the following are square numbers? 4, 6, 9, 18, 25, 81, 90, 100

2 Work these out.

  (a) 5^2 (b) 4^2 (c) 8^2 (d) 11^2

3 Find each of these.

  (a) $\sqrt{49}$ (b) $\sqrt{100}$ (c) $\sqrt{144}$ (d) $\sqrt{16}$

4 List the square numbers between 100 and 170.

5 This diagram shows a corner of a square made from 144 dots.

How many dots are along one edge?

6 A square is made from 169 dots. How many dots are along one edge?

Section F

1 Complete this list of cube numbers less than 150.   1, 8, 27, ... , ...

2 Work these out. (a) 7^3 (b) 11^3 (c) 8^3

3 Which of the following are cube numbers?   64, 25, 143, 125, 8, 9

4 (a) Find a cube number between 200 and 300.

   (b) How many cube numbers are there between 300 and 400?

5 Work out the missing numbers.

   (a) $\blacksquare^3 = 8$ (b) $\blacksquare^3 = 27$ (c) $\blacksquare^3 = 216$ (d) $\blacksquare^3 = 64$

Section G

Use the clues to find the numbers.

1
- A square number less than 100
- An odd number
- A multiple of 9

2
- A prime number
- More than 25
- Less than 30

3
- A cube number
- A multiple of 4
- Less than 10

4
- A prime number bigger than 8
- Less than 15
- A factor of 33

5
- A multiple of 5
- A multiple of 6
- Less than 35

6
- A factor of 18
- A factor of 24
- A multiple of 2
- A multiple of 3

23 Lines and angles

Section A

1 The diagram shows a rectangle made from triangles.

Say what type of triangle each one is.

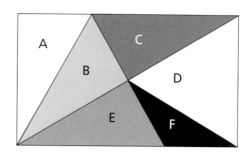

2 These diagrams show parts of the rectangle above.

Calculate the angles marked with letters in these diagrams.

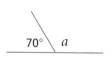 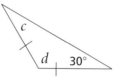 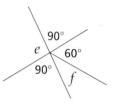

3 Calculate the angles marked with letters in these diagrams.

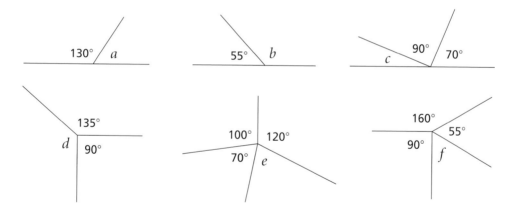

4 Calculate the angles marked with letters in these diagrams.

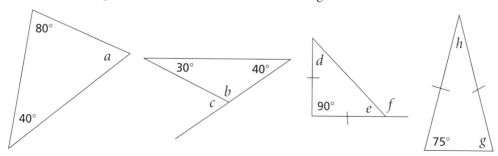

Section B

1 Calculate the angles marked with letters in these diagrams.

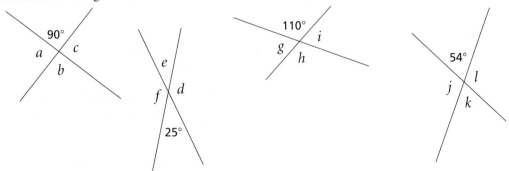

2 Calculate the angles marked with letters on this cot, fishing stool and workbench.

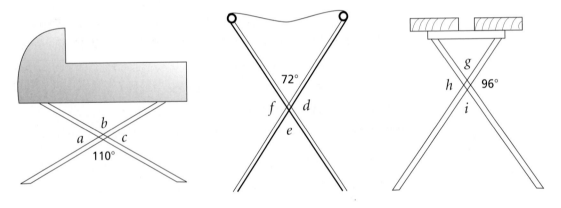

3 Calculate the angles marked with letters in these diagrams.

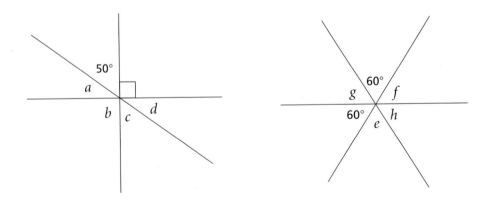

Section C

1 Find the angles marked with letters in these diagrams.

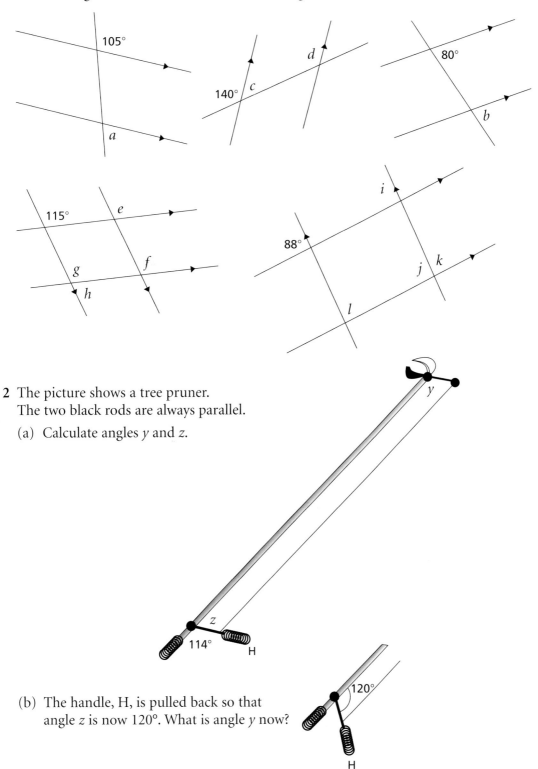

2 The picture shows a tree pruner.
  The two black rods are always parallel.

  (a) Calculate angles *y* and *z*.

  (b) The handle, H, is pulled back so that
      angle *z* is now 120°. What is angle *y* now?

65

Section D

1 Find the angles marked with letters in these diagrams.

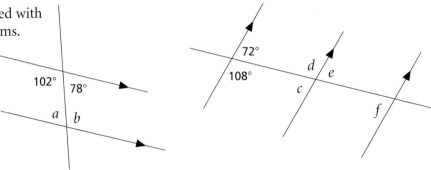

2 The picture shows an extending arm.

Copy and complete these statements.

(a) Angles a and _____ are alternate angles.

(b) Angles b and _____ are alternate angles.

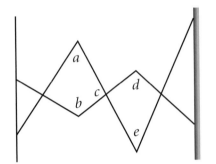

3 The pictures show a gate and a shed door. Find the angles marked with letters.

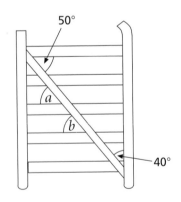

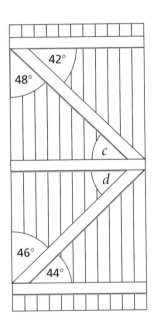

Section E

1 The sides of this car jack are parallel.

  (a) Which angle is alternate to angle *a*?

  (b) If angle *a* is 40° and angle *b* is 40°, find the unknown angles. Give your reasons.

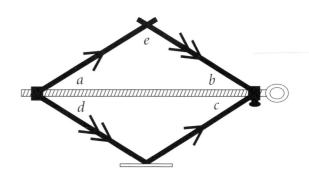

2 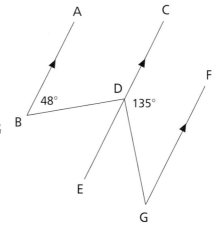 The diagram shows part of a lattice window.

  (a) Name three pairs of corresponding angles.

  (b) If angle *a* is 105°, find the unknown angles.

3 The lines AB, CE and FG are parallel.

  Find angles

  (a) BDE   (b) BDC   (c) EDG

  (d) BDG   (e) DGF

  Give reasons for each answer.

4 (a) What type of quadrilateral is BCDE?

  (b) Find angles

     (i) CBE   (ii) BED

     (iii) ABE   (iv) BEA

     (v) BAE

  Give reasons for each answer.

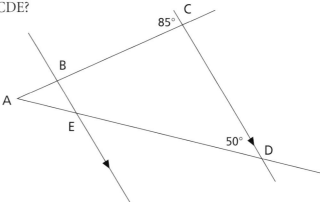

24 *Fractions*

Sections A, B and C

1 What fraction of each shape is shaded?   (a)   (b)

2 The diagrams show two equivalent fractions. What fractions are they?

3 Copy these and find the missing numbers.

   (a) $\frac{1}{2} = \frac{\blacksquare}{8}$ (b) $\frac{1}{3} = \frac{\blacksquare}{12}$ (c) $\frac{2}{5} = \frac{\blacksquare}{15}$ (d) $\frac{3}{4} = \frac{\blacksquare}{16}$

4 Write each of these fractions in its simplest form.

   (a) $\frac{6}{10}$ (b) $\frac{5}{15}$ (c) $\frac{8}{12}$ (d) $\frac{12}{20}$ (e) $\frac{9}{21}$

5 Write, in its simplest form, the fraction of the square that is shaded.

   (a) 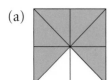   (b) 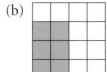   (c) 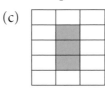   (d)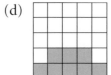

6 (a) Which fractions in this list are equivalent to $\frac{1}{3}$?

   $\frac{3}{9}$ $\frac{5}{20}$ $\frac{6}{24}$ $\frac{6}{18}$ $\frac{2}{6}$ $\frac{4}{16}$ $\frac{4}{20}$ $\frac{8}{24}$

   (b) Which fractions in the list above are equivalent to $\frac{1}{4}$?

Section D

In these questions write each fraction in its simplest form.

1 What fraction of this set of pens have their caps missing?

2 What fractions of these coins are showing 'Heads'?

3 Patrice took 24 photos on holiday.
16 of the photos were of people. The rest were of places.

What fraction of Patrice's photos were of people?

4 Six children in a class of 30 are left-handed.
What fraction of the class are left-handed?

5 What fraction of
   (a) WEDNESDAY is WED   (b) MANCHESTER is CHEST
   (c) BRONTOSAURUS is TO   (d) UNREPRESENTATIVE is SENT

6 Class 10R consists of 12 fourteen-year-olds and 20 fifteen-year-olds.
What fraction of the class are fifteen-year-olds?

7 Last season, Holby City football team played 24 matches.
They won 10, drew 8 and lost 6.
What fraction of their matches did they
   (a) win   (b) draw   (c) lose

8 Josie keeps rabbits.
She has 12 white rabbits, 10 brown rabbits and 8 black rabbits.

What fraction of her rabbits are
   (a) white   (b) brown   (c) black

Section E

1 Work out
   (a) $\frac{1}{3}$ of 18   (b) $\frac{3}{4}$ of 16   (c) $\frac{1}{5}$ of 35   (d) $\frac{3}{5}$ of 35   (e) $\frac{2}{3}$ of 36

2 Work out
   (a) $\frac{5}{8}$ of 40   (b) $\frac{3}{5}$ of 45   (c) $\frac{3}{8}$ of 160   (d) $\frac{4}{5}$ of 250   (e) $\frac{5}{6}$ of 300

3 What words do these make?
   (a) The first $\frac{1}{3}$ of CAMERA and the second $\frac{1}{2}$ of LIVE.
   (b) The last $\frac{2}{5}$ of HIPPO, the first $\frac{1}{2}$ of LITTER and the last $\frac{3}{11}$ of MATHEMATICS.

4 A packet of biscuits normally contains 36 biscuits.
A special offer gives '$\frac{1}{4}$ extra free'.
How many biscuits are in the special offer packet?

5 A company employs 48 people.
It must lose $\frac{1}{3}$ of its workforce.
How many people must it lose?

25 3-D puzzles

Sections A and B

1 (a) This shape is made from cubes.
   There are no hidden cubes.
   Draw it on triangular dotty paper.

   (b) Draw the plan view of the shape
   on squared paper.

   (c) Draw the side view of the shape
   on squared paper.

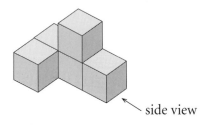

side view

2 This shape is made from cubes.
   Imagine it falls and lands on the side marked A.
   Draw the fallen shape on triangular dotty paper.

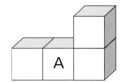

3 This diagram shows a solid object
   with some measurements.
   Use centimetre squared paper to

   (a) draw a full-scale plan view

   (b) draw a full-scale front view
   from the direction shown.

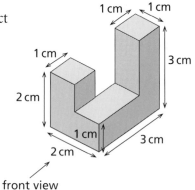

front view

4 This is a plan view of a solid built with **four** cubes.
   Draw a side view of the solid from the directions shown.

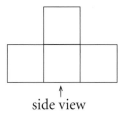

side view

5 This solid is a prism.
   (a) How many centimetre cubes
   were used to build it?

   (b) Draw a plan of the prism
   on centimetre squared paper.

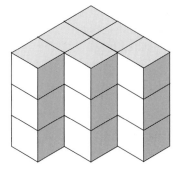

Sections C and D

1 Which of these are nets of a cube?

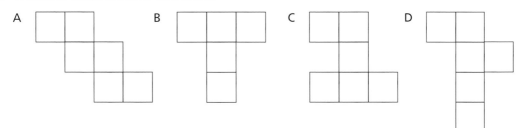

2 Imagine folding up this net to make a dice.
   (a) Which number will be opposite the 5?
   (b) Which number will be opposite the 6?

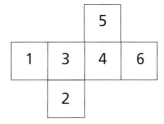

3 This is a matchbox tray with its measurements.
   (a) Draw an accurate net of the tray
       on centimetre squared paper.
   (b) What is the area of the net?

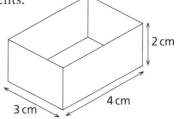

4 Here are three different
   views of the same cube.
   Which letters are opposite each other?

5 This is a triangular prism.
   The ends of the prism are isosceles triangles.
   Draw an accurate full-size net for this prism.

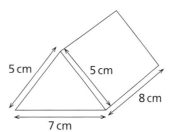

Section E

1 This shape is made from six cubes.
Which of the shapes below are
a mirror image of this shape?

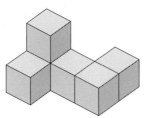

A 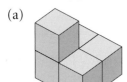 B 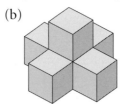 C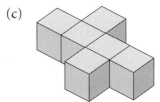

2 These shapes are made from cubes.
How many planes of symmetry does each shape have?

(a) 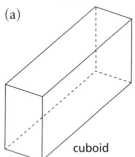 (b) 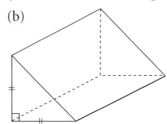 (c)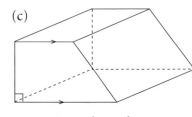

3 How many planes of symmetry does each of these shapes have?

(a)

cuboid

(b)

isosceles triangle prism

(c)

trapezium prism

26 *Written multiplying and dividing*

Sections A and B

1 Work out

(a) 38 × 5   (b) 142 × 4   (c) 215 × 8   (d) 467 × 9

2

0	1	2	3	4	5	6	7	8	9
P	A	T	C	H	W	O	R	K	S

Work out the multiplications below.
Change the digits in each answer to letters using the code above.

(a) 6 × 613   (b) 314 × 7   (c) 1852 × 5

(d) 3155 × 3   (e) 4 × 8543   (f) 12 553 × 3

3 A carton holds 9 packets of cereal.
How many packets of cereal will there be in 15 cartons?

4 (a) A mile is 1760 yards. There are 3 feet in a yard. How many feet are there in a mile?

(b) A stone is 14 pounds. How many pounds are there in 8 stones?

(c) There are 8 pints in a gallon. How many pints is 24 gallons?

5 Work out

(a) 0.6 × 2   (b) 1.5 × 4   (c) 2.1 × 3   (d) 2.5 × 4

6 Choose numbers from the loop to
make these multiplications correct.

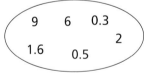

(a) 0.9 × ☐ = 1.8   (b) 8 × ☐ = 2.4   (c) 0.5 × ☐ = 4.5

(d) ☐ × 2 = 3.2   (e) ☐ × 0.4 = 2.4   (f) 4 × ☐ = 2

7 How much will it cost to buy 6 cards at £1.40 each?

8 A maths text book is 2.3 cm thick. How high will a pile of 9 of these books be?

9 Work out

(a) 1.45 × 5   (b) 2.14 × 3   (c) 4.28 × 4   (d) 6.29 × 6

10 (a) One kilogram is 2.2 pounds. How many pounds does a 7 kg bag of potatoes weigh?

(b) A litre is 1.75 pints. How many pints does a 5 litre container hold?

Section C

1 Work out

   (a) $85 \div 5$　　　(b) $76 \div 4$　　　(c) $189 \div 3$　　　(d) $836 \div 2$　　　(e) $924 \div 6$

2 Glass tumblers are sold in packs of 6.
How many packs can be made from 210 tumblers?

3 Nine classes need to share 234 books equally.
How many books will each class get?

4 Work out

   (a) $4359 \div 3$　　　(b) $2528 \div 8$　　　(c) $1015 \div 7$　　　(d) $3807 \div 9$

5 The 1650 pupils at a school are divided equally into 3 houses for sporting activities.
How many pupils are in each house?

6 Find three pairs of divisions that give the same answer.
Which division is the odd one out?

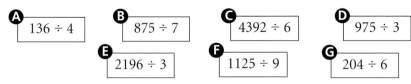

A $136 \div 4$　　**B** $875 \div 7$　　**C** $4392 \div 6$　　**D** $975 \div 3$

E $2196 \div 3$　　**F** $1125 \div 9$　　**G** $204 \div 6$

7 Choco bars are sold in packs of five.

   (a) How many packs can be made up from 84 bars?

   (b) How many bars will be left over?

8 Find three pairs which have the same remainder. Which division is the odd one out?

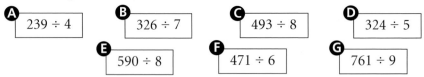

A $239 \div 4$　　**B** $326 \div 7$　　**C** $493 \div 8$　　**D** $324 \div 5$

E $590 \div 8$　　**F** $471 \div 6$　　**G** $761 \div 9$

9 Oranges are sold in bags of 5.

   (a) How many bags can be filled from 467 oranges?

   (b) How many oranges will be left over?

10 Each table in a school dining room seats 8 people.

   (a) How many tables will be needed for 675 people?

   (b) How many empty seats will there be?

11 A car transporter can carry 7 cars.
How many transporter loads will be needed to move 1580 cars?

Section D

1 Work out

(a) 8.6 ÷ 2   (b) 7.2 ÷ 4   (c) 3.6 ÷ 4   (d) 4.9 ÷ 7

(e) 1.25 ÷ 5   (f) 3.54 ÷ 3   (g) 12.35 ÷ 5   (h) 15.3 ÷ 9

2 A plank of wood measuring 3.45 m is cut into 5 equal length pieces.
How long is each piece?

3 Work out

(a) 6.3 ÷ 2   (b) 2.5 ÷ 2   (c) 6.4 ÷ 5   (d) 8.6 ÷ 4

4 Find three pairs of divisions that give the same answer.
Which division is the odd one out?

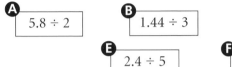

A 5.8 ÷ 2   **B** 1.44 ÷ 3   **C** 3.4 ÷ 4   **D** 14.5 ÷ 5

E 2.4 ÷ 5   **F** 3.5 ÷ 2   **G** 5.1 ÷ 6

5 A piece of cheese weighing 1.3 kg is cut into two pieces of equal weight.
How heavy is each piece?

6 A plank of wood 6 metres long is cut into 4 equal lengths.
How long is each piece?

7 15 kg from a sack of flour is divided into 4 equal bags.
How much flour is in each bag?

8 Match each division with an answer from the loop.
You can use the numbers more than once.

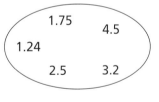

1.75   4.5   1.24   2.5   3.2

(a) 9 ÷ 2   (b) 7 ÷ 4   (c) 6.2 ÷ 5

(d) 5.25 ÷ 3   (e) 10 ÷ 4   (f) 18 ÷ 4

(g) 22.5 ÷ 5   (h) 16 ÷ 5   (i) 7.5 ÷ 3

9

A	C	E	G	H	L	P	R	Y
1.5	1.25	0.2	0.5	1.7	0.3	0.4	2.5	0.82

Find the answers to these questions. Then use the code above to change them to letters.
Rearrange each set of letters to spell a fruit.

(a) 4.1 ÷ 5   (b) 2.4 ÷ 6   (c) 10 ÷ 8   (d) 0.8 ÷ 2

     10 ÷ 4   2.7 ÷ 9   8.5 ÷ 5   7.5 ÷ 5

     5.1 ÷ 3   0.8 ÷ 4   3 ÷ 2   5 ÷ 2

     2.5 ÷ 2   2 ÷ 5   1.6 ÷ 4   4 ÷ 8

     1.6 ÷ 8   6 ÷ 4   1 ÷ 5   1.2 ÷ 6

     7.5 ÷ 3

Section E

1 Tracy was making up orange squash in a 3 litre container.
She poured in 0.45 litres of orange, and then filled the container with water.
How much water did she add?

2 Find the areas of these rectangles.

(a) 6 cm

38 cm

(b) 2.6 m

7 m

3 This pie chart shows how people
travelled to a town shopping centre.

What percentage came by train?

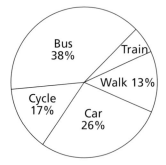

Bus 38%

Train

Walk 13%

Cycle 17%

Car 26%

4 A one pound coin has a diameter of 2.2 cm.
How long would a line of 8 one pound coins be?

5 The perimeter of a rectangle is the distance all the way round it.
This rectangle has perimeter 3.4 m.
It is 0.8 m wide. What is its length?

0.8 m

6 This square has perimeter 23 cm.
How long is each side?

7 To find the mean of a set of numbers you add them all up,
then divide by how many there are.

These are the heights of some young trees.

3.2 m   4.1 m   5.6 m   4.7 m   5.2 m

(a) What is the total of these tree heights?

(b) Find their mean height.

Mixed questions 3

1 This chart shows the ages of people using a swimming pool on a Sunday morning.

   (a) How many people aged 40 or over used the pool?

   (b) How many people altogether used the pool?

   (c) What is the modal age group of people using the pool this Sunday morning?

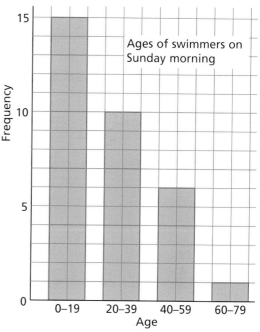

2 The table below shows the ages of people using the swimming pool on a Monday morning.

Ages of swimmers	65	66	68	46	63	66	42	60	65	56
	45	61	59	48	59	59	62	68	69	53

   (a) Copy and complete this stem and leaf table. The first two ages have been put into the table.

   (b) Write out the table again, putting the 'leaves' in order.

   (c) What is the median age of these swimmers?

   (d) What is the range of the ages of these swimmers?

   (e) Write down what you notice about the ages of people using the pool on Monday morning compared with Sunday morning.

```
0 |
1 |
2 |
3 |
4 |
5 |
6 | 5 6
Stem = 10 years
```

3 Copy the table on the right. Put ticks in the correct columns for the numbers in each row.

	is a factor of 24	is a multiple of 8	is a prime number	is a square number	is a cube number
8					
16					
27					
32					
36					
64					

4 Do these in your head, without writing down any working.

  (a) 22×4 (b) 34×5 (c) 13×4 (d) 23×5 (e) $340 \div 5$

  (f) $72 \div 4$ (g) 5×83 (h) $260 \div 5$ (i) $180 \div 4$ (j) $330 \div 5$

5 Work out each of these. Show your working clearly.

  (a) $427 + 126$ (b) $531 - 207$ (c) 54×8 (d) $2412 \div 6$

  (e) $4.3 + 2.86$ (f) $4 - 0.54$ (g) 3.75×4 (h) $6.15 \div 5$

6 Calculate the angles marked with letters in these diagrams.

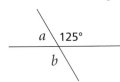 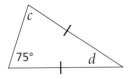 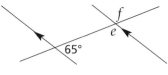

7 In the diagram, calculate the angle PBC.
   Give reasons for your answer.

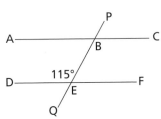

8 Write, in its simplest form, the fraction of each rectangle that is shaded.

  (a)   (b)   (c)   (d)   (e)

9 (a) Salvador has 20 rabbits. 12 of them are male.
      What fraction of Salvador's rabbits are male?
      Write the fraction in its simplest form.

  (b) Salvador also has 24 piglets.
      $\frac{3}{8}$ of them are less than 6 months old.
      How many of his piglets are less than 6 months old?

  (c) He also has 12 male pigeons and 18 female pigeons.
      What fraction of Salvador's pigeons are female?
      Write the fraction in its simplest form.

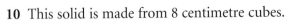

10 This solid is made from 8 centimetre cubes.

  (a) Is it a prism?

  (b) Does the solid have any planes of symmetry?
      If so how many?

  (c) On squared paper, draw a plan view,
      a front view and a side view of the solid.

Front

Side

27 Negative numbers

Section A

1 Write each list of temperatures in order, lowest first.

   (a) $^-7$°C, 3°C, $^-2$°C, 0°C, 5°C   (b) $^-1$°C, 2.5°C, $^-7.5$°C, $^-4$°C, 5°C

2 At 5 a.m. the temperature was $^-6$°C. By 9 a.m. the temperature had risen by 10 degrees. What was the temperature at 9 a.m.?

3 Work these out.

   (a) $^-3 + 9$ (b) $^-4 + 1$ (c) $^-11 + 3$ (d) $^-6 + 8$ (e) $^-5 - 2$

   (f) $^-1 - 5$ (g) $2 - 5$ (h) $4 - 7$ (i) $^-3 - 2$ (j) $^-5 + 3$

4 Work these out.

   (a) $^-3 + 8$ (b) $^-10 + 2$ (c) $2 - 9$ (d) $^-3 - 10$ (e) $2 - 12$

   (f) $^-4 - 5$ (g) $^-3 + 16$ (h) $^-20 + 5$ (i) $7 - 14$ (j) $^-12 - 4$

Section B

1 Work these out.

   (a) $3 + ^-1$ (b) $7 + ^-2$ (c) $^-4 + ^-2$ (d) $10 + ^-3$ (e) $^-1 + 9$

   (f) $^-6 + ^-2$ (g) $^-3 + ^-7$ (h) $^-2 + 11$ (i) $^-8 + ^-4$ (j) $3 + ^-9$

2 Find two numbers in the box that add up to

$$\boxed{^-6 \quad ^-4 \quad 2 \quad 5}$$

   (a) $^-10$ (b) $^-1$ (c) 1 (d) $^-4$ (e) $^-2$

3 Find two numbers in the box that add up to

$$\boxed{^-10 \quad ^-6 \quad 3 \quad 5}$$

   (a) $^-7$ (b) $^-3$ (c) $^-5$ (d) $^-1$ (e) $^-16$

4 Work these out.

   (a) $10 + ^-3 + ^-2$ (b) $^-2 + ^-3 + ^-7$ (c) $6 - 3 - 5$ (d) $^-8 + 3 + ^-2$

Section C

1 Work these out.

(a) 6 − ⁻2   (b) 5 − ⁻1   (c) 10 − ⁻3   (d) 4 − ⁻2   (e) 8 − ⁻6

2 Work these out.

(a) ⁻4 − ⁻1   (b) ⁻2 − ⁻5   (c) ⁻3 − ⁻8   (d) 3 − ⁻4   (e) 1 − ⁻1

3 Work these out.

(a) ⁻3 + 9   (b) ⁻3 − 9   (c) ⁻3 + ⁻9   (d) ⁻3 − ⁻9

(e) ⁻2 + ⁻7   (f) ⁻2 − ⁻7   (g) 2 − 7   (h) 2 − ⁻7

(i) 4 − 8   (j) ⁻4 − 8   (k) ⁻4 − ⁻8   (l) 4 − ⁻8

4 Find three pairs that give the same answer.

⁻5 + ⁻3	3 − 7	⁻4 − ⁻2	⁻10 − ⁻2	2 − 4	⁻1 + ⁻3

Section D

1 Work these out.

(a) ⁻3 × 5   (b) ⁻4 × 7   (c) ⁻2 × 8   (d) 4 × ⁻3   (e) 5 × ⁻5

(f) 10 × ⁻3   (g) ⁻4 × 25   (h) ⁻3 × 16   (i) 6 × ⁻20   (j) ⁻3 × 15

2 Work out the answers to the questions below.
Use the code to change them to letters.
Rearrange the letters to make some countries.

A	E	I	L	N	P	R	S	T	U	Y
-20	-5	1	-8	-3	-18	-6	5	-15	3	-2

(a) 6 × ⁻3
   1 − ⁻2
   ⁻2 × 3
   1 − 6

(b) ⁻5 − ⁻2
   3 + ⁻2
   ⁻2 + 7
   2 × ⁻9
   ⁻5 × 4

(c) ⁻8 − ⁻6
   2 × ⁻10
   ⁻4 × 2
   ⁻8 − 7
   ⁻3 − ⁻4

(d) 3 − ⁻2
   0 − ⁻3
   3 × ⁻2
   ⁻8 + 9
   ⁻12 + ⁻8
   8 + ⁻3

Section E

1 Work out the value of

   (a) $3a$ when $a = {}^-7$ (b) $4b$ when $b = {}^-4$ (c) $10c$ when $c = {}^-6$

   (d) $2d + 1$ when $d = {}^-4$ (e) $3e + 2$ when $e = {}^-5$ (f) $4f + 1$ when $f = {}^-5$

2 Work out the value of

   (a) $4p - 2$ when $p = {}^-3$ (b) $2q + 7$ when $q = {}^-3$ (c) $3r - 1$ when $r = {}^-4$

   (d) $2s - 6$ when $s = {}^-5$ (e) $3t + 6$ when $t = {}^-3$ (f) $4u - 3$ when $u = {}^-10$

3 Match these expressions to their answers when $p = {}^-3$.

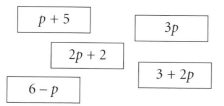 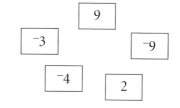

4

A	B	C	D	E	F	G	L	N	R
⁻8	4	⁻9	3	⁻6	⁻10	⁻12	⁻2	⁻17	⁻7

Find the value of each expression below when $x = {}^-5$.
Use the code to change your answers to letters.
Rearrange the letters to make a country.

$2x + 3$ $^-4 + x$ $2x$ $x + {}^-1$ $4x + 3$ $x - 3$

5 Find the value of each of these expressions when $t = {}^-3$.
Use the code in question 4 to find a letter for each answer.
Rearrange the letters to make another country.

$4t$ $t + 6$ $2t$ $4t - 5$ $2t - 2$ $^-5 - t$ $5t - 2$

28 *Fractions, decimals and percentages*

Section A

1 (a) What fraction of this shape is shaded?

   (b) Write this fraction as a percentage.

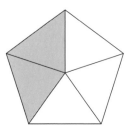

2 Write these in order of size, starting with the smallest.

   70%, $\frac{1}{2}$, 40%, $\frac{3}{5}$, 30%, $\frac{8}{10}$

3 Write these percentages as fractions, in their simplest form.

   (a) 30%   (b) 25%   (c) 60%   (d) 80%

4 Sort these into four matching pairs.

| 20% extra | $\frac{3}{5}$ more | 60% increase | $\frac{1}{5}$ more |

| 50% extra | $\frac{1}{4}$ extra | Extra $\frac{1}{2}$ | 25% more |

Section B

1 Write these decimals as percentages.

   (a) 0.3   (b) 0.6   (c) 0.9

2 Write these percentages as decimals.

   (a) 20%   (b) 75%   (c) 80%

3 Write these in order of size, starting with the smallest.

   60%, 0.75, 25%, 0.4, 50%, 0.3

4 Sort these into four matching pairs.

| 0.2 | 70% | 0.5 | 0.25 |

| 50% | 25% | 20% | 0.7 |

Sections C and D

1 Which is larger, 0.25 or $\frac{3}{10}$? Explain your answer.

2 Which is smaller, 0.4 or $\frac{1}{4}$? Explain your answer.

3 Write these in order of size, starting with the smallest.

    (a) $\frac{1}{4}$, 0.3, 20%, $\frac{2}{5}$, 10%   (b) $\frac{3}{4}$, 0.8, 70%, $\frac{6}{10}$, 90%

4

E	G	I	H	L	N	O	P	R	T	Y
$\frac{1}{2}$	20%	0.3	$\frac{1}{4}$	$\frac{2}{5}$	0.7	$\frac{3}{4}$	80%	0.6	90%	0.1

    Use this code to find a letter for each fraction, decimal or percentage below. Rearrange each set of letters to spell an animal.

    (a) 0.75, 30%, $\frac{7}{10}$, 0.4   (b) $\frac{1}{5}$, 0.9, 60%, $\frac{3}{10}$, 50%

    (c) 75%, 0.25, 70%, $\frac{3}{5}$, 30%   (d) 75%, $\frac{9}{10}$, 0.8, 70%, $\frac{1}{10}$, 25%

5 Use the code above to make up your own code problem for H I P P O.

6 Copy and complete this table.

Fraction	Decimal	Percentage
$\frac{3}{4}$		
	0.6	
$\frac{7}{10}$		
		40%
	0.9	

7 Write each of these lists in order of size, smallest first.

    (a) $\frac{4}{5}$, 0.9, $\frac{3}{4}$, 70%, $\frac{1}{2}$ (b) $\frac{3}{5}$, $\frac{1}{2}$, $\frac{3}{4}$, 40%, 0.8

8 By dividing, change these fractions into decimals.

    (a) $\frac{3}{4}$ (b) $\frac{3}{5}$ (c) $\frac{1}{8}$ (d) $\frac{7}{8}$

9 Write this list in order of size, smallest first.

       0.4, 0.6, $\frac{2}{3}$, 50%, $\frac{1}{3}$, 25%

Section E

Do all the questions in this section in your head.

1 Work out

   (a) 10% of £40   (b) 20% of £40   (c) 70% of £40

   (d) 30% of £20   (e) 40% of £50   (f) 60% of £40

2 A restaurant adds 10% to the cost of a meal, for service.
Jason's meal costs £15.

   (a) How much does the restaurant add on for service?

   (b) How much does Jason pay altogether?

3 20% of an orange's weight is peel.
If an orange weighs 80 g, how much does its peel weigh?

4 A packet of sweets normally contains 200 g.
It now contains '25% extra free'.

How many grams extra do you get free?

5 Jake is given £60 for his birthday.
He spends $\frac{1}{3}$ of this on clothes, 25% on CDs and 20% on going out.

   (a) How much does he spend on clothes?

   (b) How much does he spend on CDs?

   (c) How much does he spend on going out?

   (d) After spending all of these amounts, how much
       does he have left from his £60?

Section F

1 Copy and complete this table.

Fraction	Decimal	Percentage
$\frac{81}{100}$		
	0.01	
		87%
		6%

2 Write each of these lists in order of size , smallest first.

   (a) 0.27, $\frac{1}{4}$, $\frac{1}{5}$, 30%, 0.4   (b) 0.8, $\frac{3}{4}$, $\frac{3}{5}$, 0.67, $\frac{9}{10}$

   (c) 25%, $\frac{1}{5}$, $\frac{1}{2}$, 0.06, 0.1   (d) $\frac{4}{5}$, 0.9, $\frac{3}{4}$, 0.85, 60%

Section B

1 Work out the circumference of each of these tins, roughly.

(a) diameter 5 cm

(b) diameter 12 cm

(c) diameter 16 cm

2 This wedding cake has three layers.
A ribbon goes round the outside of each layer.

(a) The diameter of the bottom layer is 30 cm.
Roughly how long will the ribbon be?

(b) The diameter of the middle layer is 20 cm.
Roughly how long will this ribbon be?

(c) The diameter of the top layer is 15 cm.
Roughly how long will the top ribbon be?

3 ← 40 cm →

This lampshade needs a piece of tape sewn round
the top and another piece at the bottom.
Roughly how much tape will be needed altogether?
(Show your working clearly.)

← 60 cm →

4 This traffic roundabout is a circle 15 metres across.
New kerbstones are being laid around its circumference.
Each kerbstone is $\frac{1}{2}$ metre long.

Roughly how many will be needed?

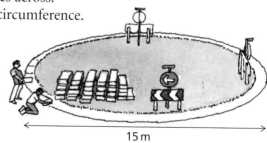

15 m

5 Cosima is putting tape around this roll of carpet.
The diameter of the roll is about 25 cm.

   (a) Roughly what length of tape will she need to
   go around the carpet once?

   (b) She puts three pieces of tape around the roll.
   About how much tape does she need altogether?

6

Ferris Wheel 1893

This is a picture of the first Big Wheel ever built.
It was 40 m from the centre to the 'cars' on
the outside.

   (a) About how high from the ground
   was the top car?

   (b) Roughly what was the circumference
   of the wheel?

Section C

1 Mary and Paul are measuring some trees.
They measure the circumference of each tree.
Then they work out roughly the diameter
and radius of each tree.

Copy and complete the table below
for the trees that they measured.

Type of tree	Circumference	Diameter	Radius
Oak	120 cm		
Silver Birch	60 cm		
Horse Chestnut	150 cm		
Yew	210 cm		
Beech	75 cm		

Section D

Use the π button on your calculator when doing these questions.
Give each answer to one decimal place.

1 Work out the circumference of each of these circles.

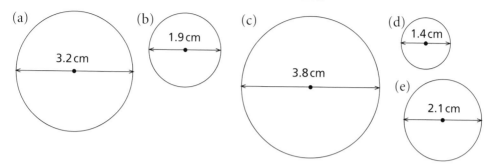

(a) 3.2 cm

(b) 1.9 cm

(c) 3.8 cm

(d) 1.4 cm

(e) 2.1 cm

2 Measure the diameter of each of these circles, and write it down. Calculate the circumference of each circle.

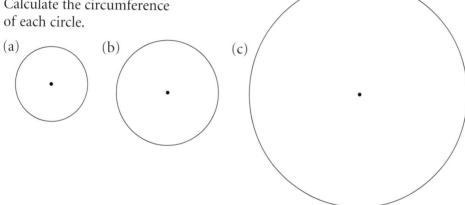

(a)　(b)　(c)

3 Harry cuts a piece of plastic 80 cm long to go round the circular end of a lampshade. The diameter of the lampshade is 26 cm.

Is the plastic long enough?
Explain your answer.

4 Bill designs labels to go round tins of food.

He cuts out a label to go round this tin. Draw a sketch of the flat label and mark the measurements.

8.5 cm

6 cm

31 *Areas of parallelograms*

Sections B and C

1 Find the area of each of these parallelograms.

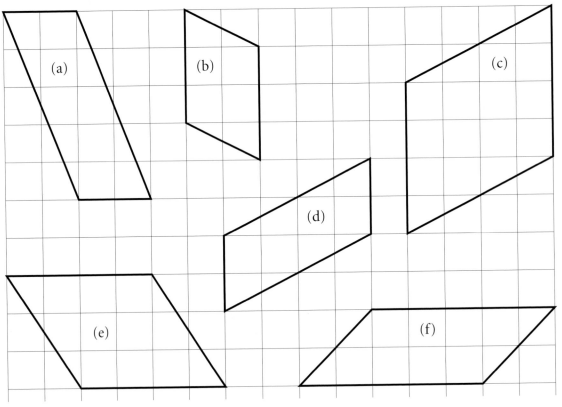

2 On centimetre squared paper, draw

  (a) three different parallelograms each with an area of $12\,\text{cm}^2$

  (b) three different parallelograms each with an area of $15\,\text{cm}^2$

3 Find the areas of these parallelograms by
  measuring the base and the height shown.

  (a)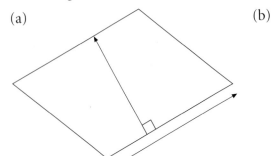

  (b)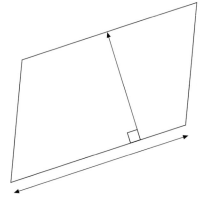

4 Find the areas of these parallelograms.
   You will not need to use all the measurements.

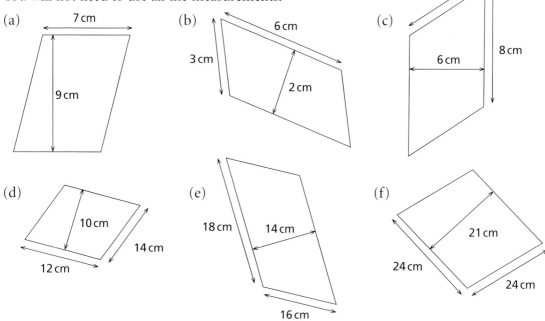

(a) 7 cm, 9 cm

(b) 6 cm, 3 cm, 2 cm

(c) 7 cm, 6 cm, 8 cm

(d) 10 cm, 14 cm, 12 cm

(e) 18 cm, 14 cm, 16 cm

(f) 21 cm, 24 cm, 24 cm

Section D

1 Find the areas of these parallelograms.

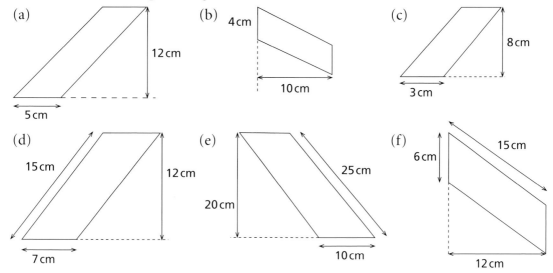

(a) 12 cm, 5 cm

(b) 4 cm, 10 cm

(c) 8 cm, 3 cm

(d) 15 cm, 12 cm, 7 cm

(e) 25 cm, 20 cm, 10 cm

(f) 6 cm, 15 cm, 12 cm

32 Gathering like terms

Section A

1 Write an expression for the perimeter of each shape.
  Write each expression as simply as possible.

(a) 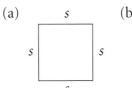 (b) 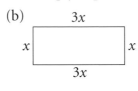 (c) 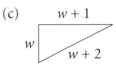 (d)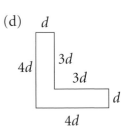

2 Simplify each of these expressions.

(a) $x + x + x$ (b) $2e + 4 + e - 1$ (c) $j + 2j + j - 6$

(d) $m + 2 + 2m - 3$ (e) $w + 2 + 2w + 5w + 8$ (f) $100w + 10w + 7$

3 This number track goes up in steps of 2.
  Write down the three missing expressions.

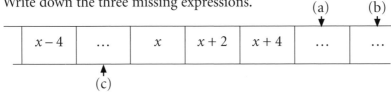

4 Look at these rods. Write an expression for each length marked **?**

(a) 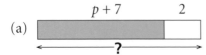 (b)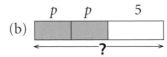

(c)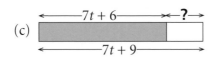

5 This is a grid of expressions.

(a) Write down the sum of the expressions in row 1.
    Write this as simply as possible.

(b) Write expressions for the sum of

   (i) row 3   (ii) column A

   (iii) column C

	A	B	C
1	$x + 3$	$x + 5$	$x + 7$
2	$2x + 3$	$2x + 5$	$2x + 7$
3	$3x + 3$	$3x + 5$	$3x + 7$

Sections B and C

1 Work out and simplify an expression for the perimeter of each of these.

(a)
    (b)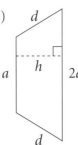
    (c)
    (d)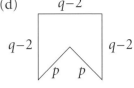

2 (a) Work out and simplify an expression for the perimeter of this rectangular tile.

(b) Two of the tiles are put together as shown.

Write an expression for the perimeter of this shape. Write your answer as simply as possible.

3 Simplify each of these expressions.

(a) $2n + 3m + m + n$ (b) $3r + s + 4r + 2s$ (c) $5p + 2q + q + 3p$

(d) $2a + 5 + 3b + a - 2$ (e) $3g + 5 + 2f + 4 + 2g + f$ (f) $7j - 3 + 2j + 4 + 3k + 2k$

4 Look at these rods.
Write an expression for each length marked **?**

(a) 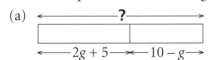   (b)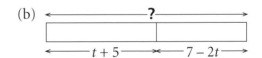

5 Sketch each diagram and write an expression for the missing angles.

(a)
right-angled triangle

(b)

(c)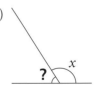
isosceles triangle

6 Simplify each of these expressions.

(a) $6a + 3b - 2a + b$ (b) $3r - 2s + r + 4s$ (c) $5c + 3d - 3c - d$

(d) $10 - 2m + n - 3m - 2n$ (e) $6p - 2q + 4 - 4p + 3q - 2$

Section D

1 A rectangle has width w cm.
  The length is 4 cm more than the width.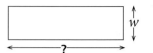

   (a) Write down an expression for the length of the rectangle.

   (b) If P is the perimeter of the rectangle,
      write down a formula for P in terms of w.

2 Andrew has x ping-pong balls in his box.
  His mum buys him four more balls.

   (a) Write an expression for the total
      number of ping-pong balls he has now.

   (b) His younger brother then takes 6 balls from the box.
      Write an expression for the number of balls in the box now.

3 Paper cups cost 12p each.

   (a) Write an expression for the cost of c cups.

   (b) Paper plates cost p pence each.
      Write an expression for the total cost of 1 cup and 1 plate.

   (c) Write an expression for the total cost of 3 cups and 3 plates.

4 (a) I have a roll of material 10 metres long.
      I cut off x metres for a costume.
      How much material is left?

   (b) I have another roll of material w metres long.
      I need to cut off material for 2 costumes
      (each needs x metres).
      How much material is left?

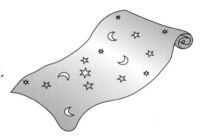

5 Damien buys x mugs at £3 each.
  He pays for them with a £20 note.
  He receives c pounds in change.
  Write down an expression for c in terms of x.

33 Connections

Section A

1 Two film critics watched some films. They gave each film a score out of 10.
The scores are shown in the table.

	Film									
---	A	B	C	D	E	F	G	H	I	J
Critic 1	6	8	2	4	4	9	7	6	7	10
Critic 2	6	7	2	6	5	10	4	8	7	8

(a) Which film did critic 1 like the least?

(b) Which was critic 2's favourite film?

(c) Draw a scatter diagram on squared paper with both scales going from 0 to 10.

(d) Do you think that the film critics like similar films?
Give a reason for your answer.

2 The table shows the measurements in centimetres of
the length and width of ten leaves taken from a tree.

	Leaves									
---	A	B	C	D	E	F	G	H	I	J
Width (cm)	1.3	2.9	2.7	1.9	3.1	2.3	2.7	3.3	2.5	1.8
Length (cm)	3.2	6.6	4.8	4.0	6.5	4.9	5.5	7.0	5.0	3.0

(a) What is the length of the longest leaf?

(b) Is the longest leaf also the widest leaf?

(c) Plot the measurements on a scatter diagram
with axes marked like this.

(d) What does the scatter diagram tell you
about the connection between
the lengths and the widths of the leaves?

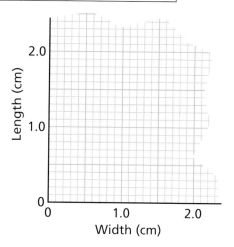

93

Section B

1 A group of students took a mathematics examination in year 6 and another mathematics examination in year 7.

Their results are shown in the table.

	Student											
	Ann	Boz	Carol	Dan	Ethan	Fiona	Gary	Hasha	Ian	Jade	Kabay	Liam
Year 6 mark (%)	93	84	62	76	55	98	66	51	88	61	58	79
Year 7 mark (%)	90	81	75	50	50	98	69	43	75	50	63	78

(a) Draw a scatter diagram on graph paper using these scales.

(b) What type of correlation does the graph show?

(c) What is the connection between the students' results in year 6 and year 7?

(d) Which student performed much better in year 6 than year 7?

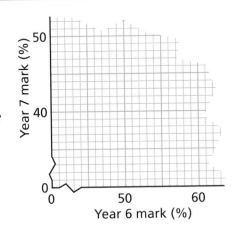

2 The table gives information taken from advertisements about the age of 8 cars and their sale price. All the cars are the same model.

Age (years)	3	9	7	6	13	10	5	2
Value (£)	6000	1000	2000	2500	500	2000	3500	6000

(a) Draw a scatter diagram on graph paper using these scales.

(b) What type of correlation does the graph show?

(c) Describe the relationship between the age and the value of the cars.

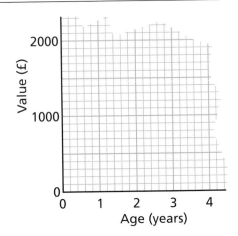

Section C

1 A magazine tested eight different makes of walking shoes.
They gave them a score based on comfort, water resistance and durability.

	Shoes							
Cost (£)	30	60	65	50	70	80	70	50
Score	15	26	33	26	40	47	35	29

(a) Show the information on a scatter diagram
   using these scales.

(b) Describe the correlation between the cost
   and the score.

(c) Draw the line of best fit.

(d) A pair of walking shoes cost £40.
   Use your graph to estimate a score.

(e) Another pair of walking shoes scored 52 marks.
   How much would you expect them to cost?

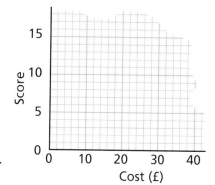

2 A football fan wanted to investigate any connection between
the number of points obtained by a football team and the
number of goals that were scored against them.

He looked at 12 teams in the league.

Points	69	65	52	44	38	55	50	44	53	36	33	24
Goals against	25	22	34	32	32	31	28	40	23	39	46	46

(a) Use the data to draw a scatter diagram on graph paper using these scales.

(b) Draw a line of best fit.

(c) What does the graph tell you about the
   connection between the points
   and goals against?

(d) Another team obtained 48 points.
   Estimate how many goals were scored
   against them.

(e) If a team had 35 goals scored against them,
   estimate how many points they obtained.

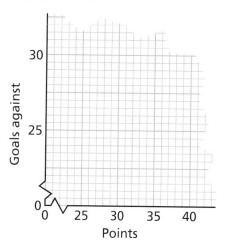

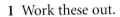

34 *Written calculation*

Show all your working clearly for all questions.

Section A

1 Work these out.

   (a) 14 × 18   (b) 23 × 41   (c) 42 × 35   (d) 37 × 27

2 Work these out.

   (a) 56 × 73   (b) 74 × 82   (c) 68 × 62   (d) 83 × 91

3 Work these out.

   (a) 175 × 21   (b) 432 × 29   (c) 721 × 92   (d) 63 × 231

4 The diagram shows a block of seats in Whyford's football stadium.

   (a) How many seats are there in the block?

   (b) There are 16 of these blocks of seats in the West Stand.
      How many seats are there in the West Stand altogether?

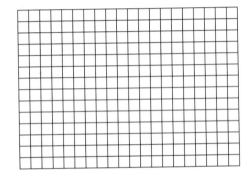

5 For an away match, Whyford supporters' club took 18 coaches of fans.
Each coach carried 41 supporters.

   (a) Do an approximate calculation to estimate how many supporters travelled by coach.

   (b) Calculate the exact number of supporters who travelled by coach.

6 The following week Whyford were playing at home.

   (a) The price of a ticket for the match was £24.
      What was the cost of 328 tickets?

   (b) The souvenir shop at the ground sold 18 car stickers at 55p each.
      How much money did they get for the car stickers?

   (c) The food kiosk sold 123 bars of chocolate costing 38p each and 247 fizzy drinks costing 44p each.

      (i) How much money did they get for the bars of chocolate?

      (ii) How much money did they get for the fizzy drinks?

Section B

1 Work these out.

  (a) 490 ÷ 14   (b) 576 ÷ 18   (c) 748 ÷ 22   (d) 384 ÷ 24

2 Work these out.

  (a) 713 ÷ 23   (b) 912 ÷ 48   (c) 901 ÷ 53   (d) 999 ÷ 37

3 Work these out.

  (a) 1794 ÷ 26   (b) 1998 ÷ 37   (c) 6080 ÷ 64   (d) 1508 ÷ 58

4 A seed merchant packs his tomato seeds into small and large packets.

  He can put 18 seeds into each small packet and 32 seeds into each large packet.
  He has 1440 seeds.

  (a) How many small packets could he make?

  (b) How many large packets could he make?

5 Cans of drink are packed in boxes of 24.
  A factory has 840 cans to pack.
  How many boxes are needed?

6 James uses pieces of string 15 cm long to tie up his rubbish bags.
  He has some string 840 cm long.

  How many rubbish bag ties can he cut from the string?

7 648 chairs are arranged in the school hall for assembly each week.
  When form 9R arranged the chairs they put 24 chairs in each row.
  Form 9S decided to put 18 chairs in each row when it was their
  turn to arrange the chairs.

  How many rows of chairs were arranged by

  (a) 9R   (b) 9S

8 A giant packet of snacks contains 12 small packets.
  The cost of the giant packet is £3.24.

  What is the cost of a small packet of snacks?

9 Oranges are sold in bags of 15 for £2.85 or bags of 12 for £2.52.

  (a) What is the price of an orange from each bag?

  (b) Are the oranges in the bigger bag cheaper and, if so, by how much each?

10 Find the lengths marked **?** for these rectangles.

(a)

? Area = 1620 cm²

45 cm

(b)

Area = 2223 cm²   57 cm

?

11 Copy and complete this division cross-number.

Across Clues

(1) 544 ÷ 34

(3) 368 ÷ 16

(4) 1107 ÷ 27

(5) 1300 ÷ 25

Down Clues

(1) 748 ÷ 44

(2) 345 ÷ 15

(3) 399 ÷ 19

(4) 924 ÷ 22

12 Margaret has 156 CDs.
She keeps them in racks.
Each rack holds a maximum of 24 CDs.

How many racks does she need?

13 A large bottle of drink holds 2500 ml of orange juice.

(a) How many glasses each holding 85 ml can be filled from the bottle?

(b) How much orange juice is left over?

14 A bag of lawn feed weighs 2 kg.
The instructions suggest it is spread at the rate of 35 g per square metre.

(a) How many grams are there in 2 kilograms?

(b) How many square metres will the bag cover?

Section C

1 4 and 5 are called a pair of factors of 20 because 4 × 5 = 20.
   31 is one factor of a pair of factors of 1457.

   Find the other factor of the pair.

2 Jamie's car travels 11 miles on every litre of petrol.
   Last month he drove 396 miles.

   (a) How many litres of petrol did he use?

   (b) If a litre of petrol costs 92p, how much did Jamie spend on petrol last month?

3 To celebrate leaving school, Sam's class went to a restaurant for a meal.
   They decided to split the cost equally between them.
   The meal cost £414.
   There were 18 people in Sam's class.

   How much did they each pay?

4 A bathroom wall is 360 cm long and 270 cm high.

   It is covered in rows of square tiles each measuring 15 cm by 15 cm.

   (a) How many rows of
        tiles are there?

   (b) How many tiles are
        there in each row?

   (c) How many tiles are
        there on the wall?

   (d) Each tile cost 36p.
        What was the total cost of the tiles?

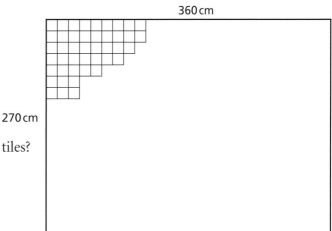

5 A drama group is putting on a play.
   The theatre they use has 52 rows of 36 seats.

   (a) Do an approximate calculation to estimate the
        total number of seats in the theatre.

   (b) Calculate the exact number of seats in the theatre.

Mixed questions 4

1 Write each of these lists in order, smallest first.

   (a) $\frac{1}{2}$, 65%, 0.25, 20%, $\frac{3}{4}$ (b) 0.15, 10%, $\frac{1}{3}$, 20%, $\frac{1}{4}$

2 Work out

   (a) 10% of 60   (b) 20% of 60   (c) 20% of 80   (d) 25% of 60   (e) 10% of 500

3 Tickleworth pensioners spent £80 on their Christmas party.
   Half of the £80 went on food, $\frac{1}{5}$ went on drink and 15% went on hiring a hall.
   The rest was spent on decorations.

   (a) How much was spent on food?

   (b) How much was spent on drink?

   (c) How much went on hiring the hall?

   (d) How much was left for decorations?

4 Write each of these lists in order, lowest first.

   (a) 2°C, ⁻3°C, 1°C, ⁻6°C, 5°C   (b) 2.5°C, ⁻1.5°C, 0°C, 3.8°C, ⁻3.2°C

5 From the loop, find two numbers that add to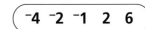

   (a) 1   (b) 2   (c) ⁻2   (d) 0

6 Work out   (a) ⁻3 × 3   (b) 3 × ⁻2   (c) 10 + 3x when x = ⁻4

7 On centimetre squared paper draw a grid with x and y going from 0 to 7.
   Then draw the parallelograms with these coordinates and find their areas.

   (a) (0, 0) (3, 0) (4, 2) (1, 2)   (b) (5, 1) (7, 3) (7, 7) (5, 5)

   (c) (0, 2) (4, 5) (4, 7) (0, 4)

8 A Christmas cake has a radius of 15 centimetres.
   Which of these measurements is closest to its circumference?

   150 cm   45 cm   30 cm   90 cm   300 cm

9 (a) A tree in Sicily had a circumference of 57 metres.
        What, roughly, was its diameter?

   (b) A bicycle wheel has a radius of 35 centimetres.
        About how many times would it go round
        if the bicycle travelled 21 metres?

10 (a) Measure the diameter of this circle in centimetres.

(b) Use the π button on your calculator (or 3.14) to work out the circumference of the circle, in cm, to one decimal place.

11 These questions were included in a survey about pocket money. Say what you think is wrong with each one.

(a) Children these days don't get enough pocket money.
Do you agree?   Yes ☐   No ☐

(b) How much pocket money do you get?

Very little ☐   Quite a lot ☐

(c) Write a better question to find out how much pocket money children get.

☒ 12 Work out the area of this parallelogram.
(You may not need to use all the measurements.)

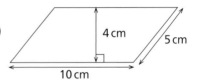

13 Write down and simplify an expression for the perimeter of each of these.

(a)

(b)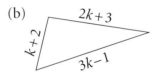

(c)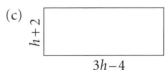

14 The scatter graph shows the length and weight of 8 eels caught in the River Severn.

(a) How long was the longest eel caught?

(b) How much did it weigh?

(c) Describe the type of correlation between the length and weight of these eels.

(d) Is the correlation strong or weak?

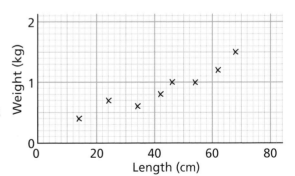

☒ 15 For each of these, first make an estimate of the answer, showing your working. Then work out the exact answer, showing clearly how you worked it out.

(a) What is the area of a rectangular floor that measures 19 m by 31 m?

(b) Jay is planting broad bean seedlings in rows.
He plants 21 seedlings in each row.
He has 819 seedlings to plant.

How many rows can he plant?

35 Graphs from rules

Section A

1 Look at the numbers in this table.
Which of the rules below is true?

x	0	1	2	3	4	5
y	0	4	8	12	16	20

$y = 2x$ $x = 4$ $y = 4x$ $x = 4y$

2 Which of the rules below is true for this table?

x	0	1	2	3	4	5
y	5	6	7	8	9	10

$y = 5x$ $y = x + 5$ $y = 5$ $x = y + 5$

3 Which rule goes with which table?

$y = 7x$ $y = 4x + 3$ $y = 2x + 4$ $y = 3x + 5$

A

x	0	1	2	3	4	5
y	4	6	8	10	12	14

B

x	0	1	2	3	4	5
y	0	7	14	21	28	35

C

x	0	1	2	3	4	5
y	5	8	11	14	17	20

D

x	0	1	2	3	4	5
y	3	7	11	15	19	23

Section B

1 (a) Copy and complete this table for the rule $y = x + 6$.

x	0	1	2	3	4	5
y						

(b) On axes like the ones shown, plot the points from the table. Join the points with a line.

(c) From the graph, what is y when $x = 3.5$?

(d) What is the value of x when $y = 7.5$?

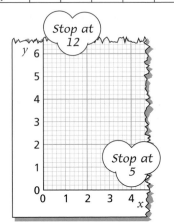

2 For the rule $y = 3x$

(a) Copy and complete this table.

x	⁻2	⁻1	0	1	2	3
y	⁻6					

(b) On axes like the ones on the right,
plot the points from the table.
Join the points with a line.

(c) Copy and complete these
coordinates of points on the line.
(0.5, …) (2.5, …) (…, ⁻4.5)

3 For the rule $y = 2x - 4$

(a) Copy and complete this table.

x	⁻2	⁻1	0	1	2	3
y	⁻8					

(b) Using the same axes you used
for question 2, plot the points.
Join them with a line.

4 Draw the graph of $y = 2x - 3$ on graph paper.
Use values of x between ⁻1 and 4.

5 On graph paper, draw $y = x + 4$
for values of x between ⁻2 and 2.

6 Draw the line $y = 3x + 2$ on graph paper.
Use values of x between ⁻1 and 3.

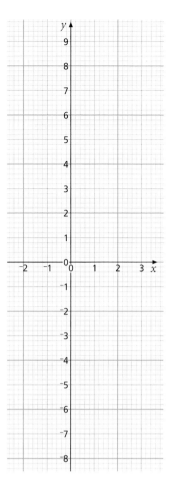

Section C

1 Copy this diagram on to squared paper.

(a) Label each line with its equation.

(b) On the same diagram draw
and label the line $y = x$.

(c) Write down the coordinates of the
points where $y = x$ crosses
each of the lines drawn.

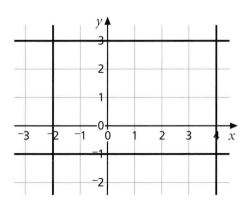

2 On squared paper, draw x and y axes both going from $^-5$ to 5.

   (a) On your axes draw and label lines with these equations.
   $$x = {}^-4, \quad y = 2, \quad x = 3, \quad y = {}^-5$$

   (b) The four lines form a shape. What is the name of the shape?

   (c) Write down the coordinates of the corners of the shape.

Section D

1 For the formula $x + y = 12$

   (a) When $x = 0$, what is y?

   (b) When $y = 0$, what is x?

   (c) Copy and complete the table.

x	0	2	4	6	8	10	12
y							

   (d) On graph paper, draw axes with both x and y going from 0 to 14.
   Plot the points from the table.
   Draw and label the line $x + y = 12$.

   (e) Copy and complete these coordinates of points on the line (4.5, ...) (..., 9.5).

2 We want to draw the graph of $x + 4y = 10$.

   (a) Copy and complete this
   table for $x + 4y = 10$.

x			
y	0	1	2

   (b) On graph paper, draw axes with x from 0 to 10 and y from 0 to 3.
   Plot the points from the table.
   Draw and label the line $x + 4y = 10$.

   (c) From the graph, what is y when $x = 0$?

   (d) What is x when $y = 1\frac{1}{2}$?

3 (a) Copy and complete this table
   of values for the rule $3x + y = 15$.

x	0	1	2	3	4	5
y						

   (b) On graph paper, draw an x-axis going from 0 to 6 and y-axis from 0 to 16.
   Draw and label the line $3x + y = 15$.

36 Chance

Section A

1 On a cube 3 faces are painted green, 2 are painted red and the other face is painted white.
The cube is rolled and the colour of the top face noted.

What is the probability that the top face is

(a) red   (b) white   (c) green   (d) not red   (e) yellow

2 Seven balls numbered 1 to 7 are placed in a lottery machine.
One ball is released.

What is the probability that the number on it is

(a) 5   (b) not 6   (c) even

(d) odd   (e) smaller than 3   (f) 8

3 8 blue cubes and 5 red cubes are put in a box.

(a) How many cubes are there in the box?

You take one of the cubes out of the box without looking.
What is the probability that the cube is

(b) blue   (c) red

4 A bag contains 3 dark chocolate cookies, 5 raisin cookies and
4 white chocolate cookies.
You take a cookie from the bag without looking.

What is the probability that the cookie is

(a) a dark chocolate cookie   (b) a raisin cookie

(c) a chocolate cookie   (d) not a white chocolate cookie

5 These 7 cards have pictures of black, grey and white shapes on one side.
The cards are shuffled and placed picture side down.

A card is turned over at random.

What is the probability that it is

(a) a black shape

(b) a shape with 4 sides

(c) not a white shape

(d) a circle

(e) not a square

Section B

1 Jim has 3 shirts and 2 ties which he wears to work.
He has a white shirt, a grey shirt and a pink shirt.

One of his ties is red and the other is mauve.

Shirt	Tie
W	R
W	M

(a) Copy and complete this list of all the possible
combinations of shirt and tie that Jim could wear to work.

(b) How many different possible combinations are there?

2 For breakfast Jim eats cereal, and then toast with marmalade.
He can choose from 4 different cereals:
cornflakes, wheat biscuits, rice crunchies or porridge.

He can choose between orange, lemon or tangerine marmalade.

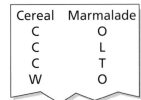

Cereal	Marmalade
C	O
C	L
C	T
W	O

(a) (i) Copy and complete this list of all the possible
combinations of cereals and types of marmalade
he can eat for breakfast.

(ii) How many possible combinations of cereal and
marmalade are there?

(b) He discovers he has run out of porridge.
How many possible combinations of cereal and marmalade are there now?

3 Jim buys a sandwich for lunch from
Jane's Sandwich Bar.

The sandwich bar uses 4 different kinds of bread –
white, brown, granary and harvestgrain.
For fillings, Jane uses ham, cheese, prawns or egg.

> JANE'S SANDWICH BAR
> **We sell over
> 20 different sandwiches.**

(a) Make a list of all the different possible sandwich combinations.

(b) How many different sandwiches are available?

(c) Does the sandwich bar really sell more than 20 different kinds?

Section C

1 The arrows on these two spinners are spun.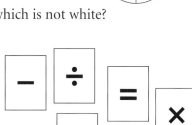

(a) List all the possible outcomes.

(b) How many outcomes are there altogether?

(c) What is the probability of getting black and 2?

(d) What is the probability of getting grey and an even number?

(e) What is the probability of getting 3 and a colour which is not white?

2 Five cards have the symbols −, =, +, × and ÷ on them.
They are shuffled and placed face down.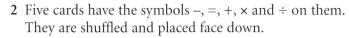

A card is chosen at random and the symbol noted.
The card is replaced and they are shuffled again.
A second card is chosen at random and recorded.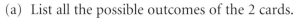

(a) List all the possible outcomes of the 2 cards.

(b) How many outcomes are there altogether?

(c) What is the probability that both cards are the same?

(d) What is the probability that one card is + and the other card is − ?

3 Two fair spinners with crown, star, diamond and pearl symbols are spun.

(a) Make a list to show all the possible outcomes.

(b) How many outcomes are there altogether?

(c) What is the probability that

(i) both spinners show a crown

(ii) both spinners show the same symbol

(iii) both spinners show different symbols

4 Eight cards are put in 3 piles as
shown in the picture.

The cards are placed in their piles
face down and shuffled.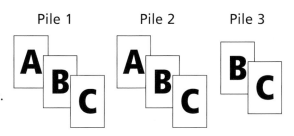

Pile 1   Pile 2   Pile 3

A card is chosen at random from each pile.

(a) List all the possible outcomes.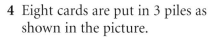

(b) What is the probability that

(i) the cards all show the same letter   (ii) the cards have exactly 2 letters the same

(iii) the letters are all different   (iv) there is exactly one A

(v) there are exactly two As

Section D

1 A dice numbered 1, 2, 3, 4, 5, 6 and a dice numbered 1, 1, 3, 3, 5, 6 are rolled.
The numbers showing on the dice are added together.

(a) Copy and complete this table showing all the possible totals.

(b) What is the probability of getting a total of 2?

(c) What is the probability of scoring more than 6?

(d) Which is the most likely total?

(e) What is the probability of this total?

First dice

+	1	2	3	4	5	6
1	2					
1						
3		5				
3						
5						
6						

Second dice

2 Matthew spins a 5-sided spinner numbered 1, 2, 3, 4, 5 and throws an ordinary dice.

He finds the **difference** in the numbers to get his score.

(a) Copy and complete the table to show all the possible scores.

(b) What is the probability of scoring 5?

(c) What is the probability that the score is greater than 3?

(d) What score is most likely? What is the probability of this score?

(e) What is the probability of scoring either 1 or 2?

Spinner

−	1	2	3	4	5
1				3	
2					
3		1			
4					
5					
6					

Dice

3 A 5-sided spinner numbered 1, 1, 2, 3, 4 and another numbered 1, 2, 3, 4, 4 are spun.

The numbers are **multiplied** together to get a score.

(a) Copy and complete this table.

(b) What is the probability that the score is

(i) less than 4

(ii) an even number

(iii) greater than 10

First spinner

×	1	1	2	3	4
1					
2					
3					
4			8		
4					

Second spinner

37 *Rounding with significant figures*

Section A

1 Round these numbers to one significant figure.

   (a) 621   (b) 377   (c) 2399   (d) 78   (e) 4062

   (f) 5920   (g) 64   (h) 85 195   (i) 307   (j) 9456

2 Write a headline for each of these stories. Round each number to one significant figure.

   (a) The Jackpot this week is £3 135 426.

   (b) 47 186 people say no to a new airport runway.

3 Work out

   (a) 20×60 (b) 30×40 (c) 50×300

4 Work out a rough estimate for each of these, by rounding
   the numbers to one significant figure.

   (a) 48×19 (b) 72×31 (c) 184×22 (d) 58×285 (e) 108×386

5 A gardener needs 32 fence panels at £18 each.
   Estimate roughly how much it will cost him.

Section B

1 Round these numbers to one significant figure.

   (a) 56.2   (b) 4.83   (c) 0.512   (d) 86.45   (e) 8.17

   (f) 0.763   (g) 0.445   (h) 0.0843   (i) 0.0674   (j) 0.005 28

2 Round these numbers to one significant figure.

   (a) 4672   (b) 148.7   (c) 0.156   (d) 0.3089   (e) 0.075

3 Rewrite each of these sentences as an approximation, rounding any numbers to
   one significant figure. The first one is done as an example.

   (a) The longest cable suspension bridge has a main span of 1991 metres. (It is in Japan.)
        The longest cable suspension bridge has a main span of about 2000 metres.

   (b) The world's largest pancake was 49 feet in diameter and weighed 2.95 tons.
        (It was made in Rochdale in 1994.)

   (c) In 1999 a pumpkin was grown that weighed 513 kg.

   (d) The largest pyramid of champagne glasses contained 30 856 glasses.

   (e) The world's smallest spider has a length of 0.43 mm.

Section C

1 Rectangle P shows that $0.7 \times 0.2 = 0.14$.

   (a) What does rectangle Q show?

   (b) What does rectangle R show?

   (c) What does rectangle S show?

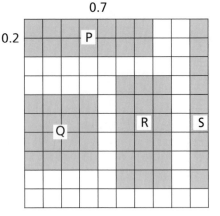

2 Copy these and fill in the missing numbers.

   (a)
   > $3 \times 4 = 12$
   > $3 \times 0.4 =$
   > $0.3 \times 0.4 =$

   (b)
   > $8 \times 4 = 32$
   > $8 \times 0.4 =$
   > $0.8 \times 0.4 =$

   (c)
   > $4 \times 5 = 20$
   > $4 \times 0.5 =$
   > $0.4 \times 0.5 =$

3 Work these out.

   (a) 0.2×0.8　　　　(b) 0.6×0.1　　　　(c) 0.5×0.9

4 (a) Write down the answer to 2×12.

   (b) Now write down the answer to each of these.

   (i) 2×1.2　　(ii) 2×0.12　　(iii) 0.2×12　　(iv) 0.2×1.2

5 (a) Write down the answer to 3×22.

   (b) Now write down the answer to each of these.

   (i) 0.3×22　　(ii) 0.3×2.2　　(iii) 0.03×22　　(iv) 0.3×0.22

6 You are told that $15 \times 19 = 285$.
   Write down the answer to each of these.

   (a) 15×0.19　　(b) 1.5×1.9　　(c) 0.15×1.9　　(d) 0.15×0.19

7 You are told that $34 \times 45 = 1530$.
   Write down the answer to each of these.

   (a) 3.4×45　　(b) 34×0.45　　(c) 3.4×4.5　　(d) 0.034×45

8 Work these out.

   (a) 30×0.5　　(b) 400×0.2　　(c) 0.6×70　　(d) 0.8×500

   (e) 0.6×0.6　　(f) 0.4×90　　(g) 0.5×60　　(h) 900×0.9

110

Section D

1   Work out a rough estimate for each of these, by first rounding each number to one significant figure.

(a) 3.12×0.625 (b) 0.723×19.5 (c) 0.294×31.6 (d) 58.7×0.542

2   52 people are going by coach to a theme park. They each pay £21.50.

(a) Estimate roughly the total cost of the trip.

(b) Is your rough estimate bigger or smaller than the exact amount?
    How can you tell without working out the exact amount?

3   Jake measures the length and width of a rectangular room.
    He says the length is 6.82 metres and the width is 3.85 metres.

(a) Estimate roughly the area of the floor of the room.

(b) Is your rough estimate smaller or larger than the actual area?

4   Kate buys a multipack of cat food containing 48 tins, each weighing 0.454 kg.
    Estimate roughly the weight of the multipack.

5   Work out a rough estimate for each of these.

(a) 8.23×0.287 (b) 0.645×31.5 (c) 4.89×0.032 (d) 68.2×0.184

(e) 0.067×31.4 (f) 318.5×0.378 (g) 0.106×21.3 (h) 0.0236×99.9

6   Match up these calculations with the rough estimates of their answers.

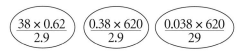

7   Work out a rough estimate for each of these.

(a) $\dfrac{38.8 \times 0.324}{2.96}$ (b) $\dfrac{53.6 \times 0.763}{3.84}$ (c) $\dfrac{213.3 \times 0.603}{27.88}$

(d) $\dfrac{0.913 \times 48.08}{3.42}$ (e) $\dfrac{0.689 \times 60.8}{5.74}$ (f) $\dfrac{0.0126 \times 768}{37.4}$

Section E

1 (a) Round 51.542 to 1 d.p.   (b) Round 6.746 to 2 d.p.
  (c) Round 8.355 to 2 d.p.   (d) Round 24.078 to 1 d.p.
  (e) Round 12.6075 to 2 d.p.   (f) Round 0.0589 to 2 d.p.

2 Use a calculator to work these out.
  (a) 2.51 × 3.45, answer to 2 d.p.   (b) 0.456 × 23.78, answer to 1 d.p.
  (c) 4.235 × 8.76, answer to 2 d.p. (d) 0.0684 × 1.27, answer to 3 d.p.

3 Polly buys 10.5 metres of ribbon costing £0.93 per metre.
  (a) Work out a rough estimate of the total cost.
  (b) Use a calculator to find the total cost. Round it to the nearest penny.

4 (a) Work out a rough estimate of the area of this poster in cm².
  (b) Use a calculator to find the area of the poster.
      Give your result to one decimal place.

29.7 cm

20.8 cm

5 One pint = 0.57 litres. Change 16 pints to litres,
  giving your answer to one decimal place.

6 A stone in weight = 6.35 kg. Charlie weighs 8.5 stones.
  How many kilograms is this?
  Round your answer to one decimal place.

Section F

Give the answers to these questions to two decimal places.

1 In October 2001, £1 was worth 1.48 US dollars (US$).
  (a) Change £64 to US dollars.   (b) Change 200 US dollars to pounds.

2 In October 2001, £1 was worth 178.06 Japanese yen.
  (a) Change £25 to yen.   (b) Change 6500 yen to pounds.

3 Angie changed £355 into Australian dollars at a rate of A$2.30 to the pound.
  How many Australian dollars did she get?

4 Peta went to Norway for a holiday.
  She had 220 Norwegian kroner left at the end of the holiday,
  and changed it into pounds at the rate of 12.4 kroner to the pound.
  How much did she get?

Sections B and C

1 Calculate the angles marked with letters in these triangles.

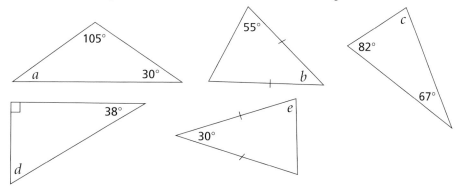

2 Calculate the angles marked with letters in these diagrams.

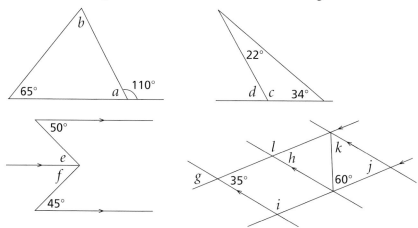

3 Calculate the angles marked with letters in these diagrams.

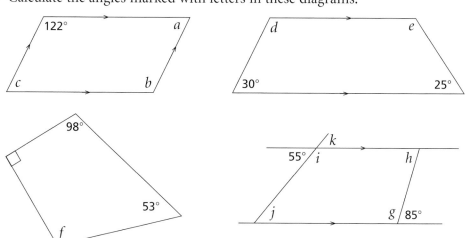

Sections D and E

1 Calculate the angles at the centre of these regular polygons.

(a)

(b)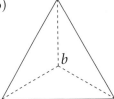

(c)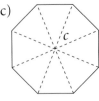

2 (a) Which of these rules describes how to find the angle at the centre
of a regular polygon if you know the number of sides?

| angle at centre = 360° – number of sides |

| angle at centre = 360° + number of sides |

| angle at centre = 360° × number of sides |

| angle at centre = 360° ÷ number of sides |

(b) Use your rule to calculate the angle at the centre of a decagon (10 sides).

3 Find the angles marked with letters on these polygons.

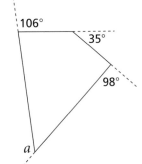

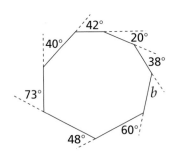

4 What size is an exterior angle on each of these shapes?

(a) a square   (b) a regular hexagon   (c) a regular decagon

5 This shape is a regular pentagon.
Point O is at the centre.

Find the angles marked with letters.

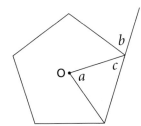

Section F

1 Copy and complete this flow diagram to find
the sum of the interior angles in a polygon.

number of sides ⟶ [− 2] ⟶ [× ...] ⟶ sum of interior angles

2 Calculate the **sum** of the interior angles in these shapes.

   (a) a pentagon   (b) a heptagon (7 sides)

3 Find the angles marked with letters in these diagrams.

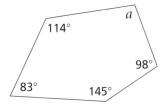
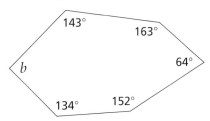

4 The sum of the angles in a nonagon (9 sides) is 1260°.
What is the size of an interior angle in a regular nonagon?

5 Calculate the size of an interior angle in these regular shapes.

   (a) a hexagon   (b) an octagon   (c) a decagon

6 In this pattern two regular pentagons
and a third regular shape meet at a point.

   (a) Find the angles p and q.

   (b) Use your answers to question 5 to
   say what the third shape is.

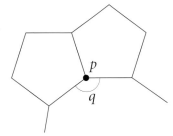

7 At each point in this pattern a square,
a regular hexagon and a regular
dodecagon (12 sides) meet.

   (a) What is the internal angle of a
   regular hexagon?

   (b) What is the internal angle of a
   regular dodecagon?

   (c) Show that the sum of the angles
   at each point is 360°.

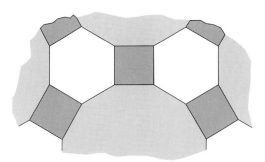

Section G

1 This pattern is made from four congruent (identical) rhombuses.

Find the angles marked with letters.

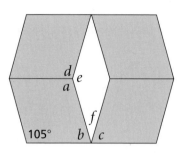

2 This diagram shows a regular octagon split into four shapes.

(a) What is the special name of quadrilateral A?

(b) What is the special name of quadrilateral B?

(c) What is the special name of triangle D?

(d) Calculate angle p, the interior angle of a regular octagon.

(e) Find angle q.

(f) Find angle r.

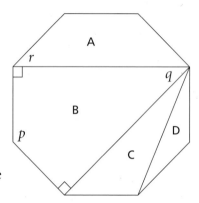

3 This pattern uses five regular pentagons.

(a) What special type of triangle is triangle ABC?

(b) Find the size of the exterior angle of a regular pentagon.

(c) What is the interior angle of a regular pentagon?

(d) Calculate angle BAC.

(e) Find angle CBA.

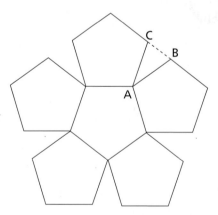

39 Ratio and proportion

Section A

Potted Prawn Pâté
(serves 4)

200 g peeled prawns
80 g butter
10 ml lemon juice
20 ml chopped parsley

Blue Cheese Dip
(serves 6)

120 ml soured cream
150 g blue Stilton cheese

Fruit Cake made with oil
(serves 8)

200 g plain flour
8 ml baking powder
400 g mixed dried fruit
20 ml sunflower oil

Chocolate Ginger Flan
(serves 10)

200 g low fat soft cheese
175 g ginger biscuits
50 g flour
75 g butter
30 ml cocoa powder
15 ml stem ginger syrup

1 How much butter would you need to make potted prawn pâté for 8 people?

2 How much of these ingredients would you need?

   (a) Soured cream to make blue cheese dip for 12

   (b) Flour to make chocolate ginger flan for 5

   (c) Lemon juice to make potted prawn pâté for 12

   (d) Mixed dried fruit to make fruit cake for 16

   (e) Blue Stilton cheese to make blue cheese dip for 24

3 (a) List the ingredients needed to make potted prawn pâté for one person.

   (b) Use your answer to list the ingredients needed
       to make potted prawn pâté for

       (i) 6 people   (ii) 10 people

4 (a) How much low fat soft cheese do you need to make
       chocolate ginger flan for one person?

   (b) Use this to work out how much low fat soft cheese you would need
       to make chocolate ginger flan for

       (i) 4 people   (ii) 12 people   (iii) 25 people

5 List the ingredients needed to make a fruit cake for 12 people.

Sections B and C

1 In each of the following examples

• find the cost of buying the items in multipacks separately

• decide which multipack gives you more for your money

(a)

(b)

2 A 50 g bar of chocolate costs 39p.
A 250 g bar of chocolate costs £1.99.

Which size is better value for money?

3 A 0.5 litre packet of washing liquid costs £1.49.
A 1.5 litre bottle of washing liquid costs £4.49.

Which is better value for money, the packet or the bottle?

4 A 500 g box of cat biscuits costs £1.86.
A 2 kg bag of cat biscuits costs £7.35.

(a) How many boxes would you need to make the same weight as a bag?

(b) How much would these boxes cost?

(c) Do you save money by buying the bag?

5 On a school trip, 80 children filled 2 coaches.

(a) How many children filled 1 coach?

(b) How many children would fill 5 coaches?

6 There are 15 apricot halves in a 500 g bag of mixed dried fruit.
How many apricot halves would you expect there to be in a 2 kg bag?

7 Frank drove 160 km travelling to work over a 4-week period.

At the same rate, how many kilometres would he drive
travelling to work over a 9-week period?

Section D

1 Natalie was going on holiday and wanted to buy some films.

Two different shops, Snappers and Quikpix were doing different offers.

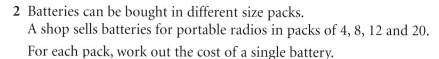

(a) How much would a single film cost at Snappers?

(b) At Quikpix, what would be the the cost of a single film?

(c) Which shop gives more films for your money?

2 Batteries can be bought in different size packs.
A shop sells batteries for portable radios in packs of 4, 8, 12 and 20.

For each pack, work out the cost of a single battery.

(a)

(b) Which pack is the best value for money?

(c) Which pack is the worst value for money?

3 In the following, use unit costs to work out which is cheaper.

(a) 5 kg for £1.35 or 4 kg for £1.12

(b) 9 kg for £27 or 2 kg for £5.80

(c) 8 litres for £4 or 10 litres for £5.50

(d) 7 metres for £8.75 or 8 metres for £9.60

4 Find the unit cost of these to the nearest penny.

(a) 3 metres of wood costing £7.99

(b) 6 kg of peanuts costing £3.99

(c) 1.5 litres of orange juice costing £1.99

(d) 25 kg of sand costing £2.99

5 In the following, find the cost of 100 ml to work out which is better value for money.

(a) 300 ml for £1.56 or 500 ml for £2.45

(b) 200 ml for £2.52 or 700 ml for £9.10

(c) 400 ml for £25.20 or 500 ml for £32

(d) 300 ml for £2.64 or 800 ml for £6.96

6 Forever aftershave is sold in different sizes.

   Large: 125 ml for £23.80   Standard: 75 ml for £14.30

   (a) Find the cost of 25 ml of both sizes, giving your answers to the nearest penny.

   (b) Which size gives better value for money?

Section E

1 Find the ratio of stars to circles in each of these pictures.

   In which pictures is this ratio 3 : 1 ?

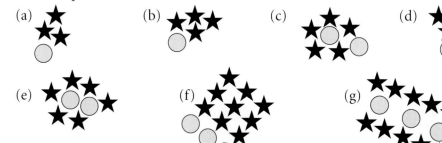

2 The ratio of apples to cherries in a pie mix is 1 : 5 .

   (a) How many cherries would you need to go with

      (i) 3 apples   (ii) 5 apples   (iii) 20 apples

   (b) How many apples would you need to go with

      (i) 10 cherries   (ii) 20 cherries   (iii) 50 cherries

3 First class and standard class seats on a train are in the ratio 1 : 10 .

   (a) How many standard class seats are there if the train has

      (i) 40 first class seats   (ii) 65 first class seats

   (b) How many first class seats are there if the train has

      (i) 300 standard class seats   (ii) 580 standard class seats

4 To make a pineapple and lime fruit drink, Amy uses pineapple juice
and lime cordial in the ratio 4 : 1 .

   (a) If she uses 600 ml of pineapple juice, how much lime cordial will she need?

   (b) If she uses 250 ml of lime cordial, how much pineapple juice will she need?

   (c) How much pineapple juice would she need if she used a glassful of lime cordial?

5 To make orange and lemon drink you need oranges and lemons in the ratio 5 : 2 .

   (a) How many oranges would you need if you used

      (i) 4 lemons   (ii) 8 lemons   (iii) 1 lemon

   (b) How many lemons would you need if you used

      (i) 15 oranges   (ii) 25 oranges   (iii) 100 oranges

Section G

1 A box contains orange and lemon slices in the ratio $2:5$.
There are 35 slices in the box.

How many orange slices are there in the box?

2 Two people share some money.
How much do they each get if they share
 (a) £20 in the ratio $2:3$ (b) £60 in the ratio $1:4$
 (c) £4.50 in the ratio $5:4$ (d) £400 in the ratio $3:5$

3 The ratio of male to female shoppers in a large department store was $2:5$.
If there were 350 shoppers in the store, how many of them were male?

4 For a football team, the ratio of goals scored at home to goals scored
at away games was $3:2$.

If the team scored 45 goals, how many did they score at home?

5 Melissa checked a box of apples to see if any were going bad.
She found that the ratio of good to bad apples was $9:1$.

If there were 80 apples in the box, how many of the apples were going bad?

6 Ken made his own mix of bird seed by mixing wheat and sunflower
seeds by weight in the ratio $3:5$.

What quantities of the two ingredients did he use if his mixture weighed
 (a) 800 g   (b) 2 kg

7 In a competition, the money awarded for the first prize and the second prize
was in the ratio $4:1$.

If the money awarded for the prizes was £100, how much was given as the first prize?

8 For a buffet party, a host prepared ham, turkey and chicken by weight
in the ratio $2:4:3$.

Find the quantity of each of the three meats prepared by the host
if the total quantity of meat prepared was
 (a) 4.5 kg   (b) 1.35 kg

9 A fertiliser contains ammonia, lime and potash by weight in the ratio $1:3:2$.
Using this ratio find the weights needed
 (a) of lime and potash with 400 g of ammonia
 (b) of ammonia and lime with 4 kg of potash
 (c) of ammonia and lime with 1 kg of potash
 (d) of ammonia, lime and potash to make 1.5 kg of fertiliser

40 *Areas of triangles*

Section C

1 Find the areas of these triangles.

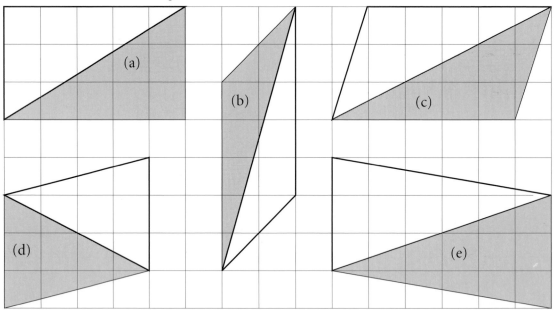

2 Find the areas of these triangles.

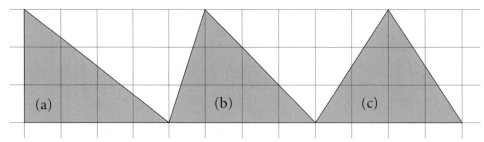

3 Copy this diagram on to centimetre squared paper.

(a) What is the area of the shaded triangle?

(b) On your diagram, draw two different
     triangles with the same area.

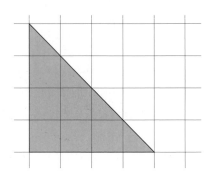

4 Find the areas of these triangles.

(a)

6 cm

7 cm

(b)

11 cm

8 cm

9 cm

(c)

18 cm

12 cm

(d)

16 cm

48 cm

20 cm

(e)

7.5 cm

7.6 cm

(f)

8.7 cm

9.2 cm

4 cm

(g)

15.4 cm

12 cm

5 (a) Find the areas of triangles A, B, C, D, E and F in this diagram.

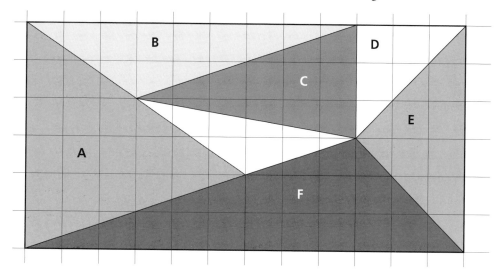

(b) What is the area of the rectangle around the diagram?

(c) Add the areas of A to F and use this to find the area of the triangle with no letter.

Sections D and E

1 Calculate the area of each of these kites.

(a)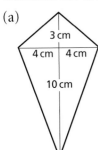

(b)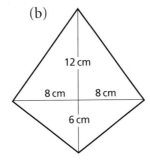

2 Find the area of these shapes.

(a)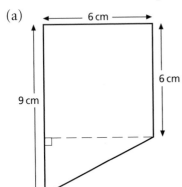

(b)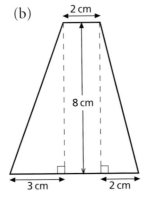

3 Calculate the area of each of these shapes.

(a)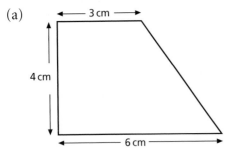

(b)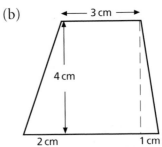

4 These are some parts for a model house.
They are cut from a rectangle of wood.

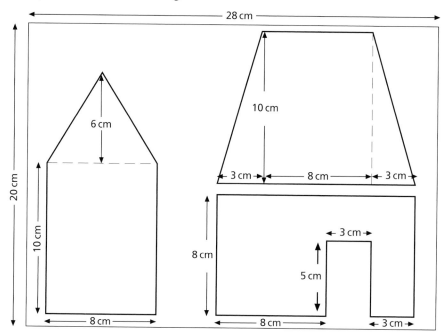

(a) What is the area of the rectangle?

(b) Calculate the area of pieces G, H and I.

(c) What is the area of wood wasted?

5 Change these areas in m² to cm².

    (a) 8 m²   (b) 50 m²   (c) 100 m²   (d) 4.5 m²

    (e) 3.6 m²   (f) 0.5 m²   (g) 0.75 m²   (h) 0.01 m²

6 Change these areas in cm² to m².

    (a) 60 000 cm²   (b) 150 000 cm²   (c) 2 000 000 cm²

    (d) 25 000 cm²   (e) 6000 cm²   (f) 500 cm²

7 Find the area of this triangle in

    (a) cm²   (b) m²

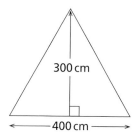

41 Trial and improvement

Sections A and B

1 The length of a rectangle is 1 m longer than its width.
   The area of the rectangle is 240 m². Find the length and the width.

2 Find the missing digits to make these calculations correct.

   (a) $18 \times 2\blacksquare = 4\blacksquare6$ (b) $966 \div 2\blacksquare = \blacksquare2$ (c) $5\blacksquare \times 48 = 27\blacksquare4$

3 Copy and complete this table square
   by placing the numbers 37, 39, 41
   and 43 in the correct places.

×		
	1591	1443
	1763	1599

4 When two numbers are added together, the answer is 60.
   When the same two numbers are multiplied together, the answer is 864.
   What are the two numbers?

5 The area of a rectangle is 120 cm² and its perimeter is 44 cm.
   What are the dimensions of the rectangle?

Section C

1 Work these out.

   (a) 13^2 (b) 8^3 (c) 15^2 (d) $12^2 - 11^2$

2 Find two square numbers between 250 and 300.

3 21^2 is 441. What is the next square number after 441?

4 A piece of square card has side length 24 cm.

   (a) What is the area of the card?

   (b) Another square card has area 729 cm².
       What is the side length of the square?

5 Find a cube number number between 300 and 400.

6 Find x if   (a) $x^3 = 2197$ (b) $x^3 = 6859$

7 The volume of this cube is 512 cm³.

   (a) How long is each edge?

   (b) What is the area of the shaded face?

Section D

1 Find a pair of consecutive numbers that add together to make 127.

2 Find three consecutive numbers that add together to make

    (a) 60   (b) 87   (c) 330   (d) 4260

3 Find a pair of consecutive numbers that **multiply** together to make

    (a) 272   (b) 650   (c) 2070   (d) 6006

4 (a) Find a pair of consecutive numbers that multiply to make 210.

    (b) Find three consecutive numbers that multiply to make 210.

5 Find a pair of consecutive **even** numbers that multiply together to make 2024.

6 Find two consecutive **odd** numbers that multiply together to make 5475.

Section E

1 Two numbers differ by 6 and multiply to give 1672. Copy and complete this table to find the two numbers.

First number	Second number	Multiply together	Result too small	too big
20	26	520	✓	
30	36	1080	✓	
39	45	1755		✓

2 Two numbers, one number double the other, multiply to make 4418. Find the two numbers.

3 Two numbers differ by 0.5 and multiply to make 264. Find the two numbers.

4 Two numbers add to make 17 and multiply to make 61.36. Find the two numbers.

5 Two numbers add to make 100 and multiply to make 2338.71. Find the two numbers.

Section F

1 The area of a rectangle is $60\,cm^2$. The length is 3 cm longer than the width.

width + 3

width

Copy and complete this table to find the width to one decimal place.

Width (cm)	Length (cm) (width + 3)	Area (cm^2)	Result too small	too big
4	7	28	✓	
5				

2 A rectangle has an area of 100 cm² and its length is 3 times its width. Find the width of the rectangle correct to one decimal place.

3 The volume of a cube is 100 cm³.

Copy and complete this table to find the length of an edge of the cube correct to one decimal place.

Length of edge (cm)	Volume (cm³)	Result	
		too small	too big
5	125		✓

Section G

Decide on your own method to solve these problems.

1 Find the missing numbers in these statements

(a) 19 × ■ = 722

(b) ■ × 36 = 1692

2 Two sticks placed end to end measure 69 cm.

One of the sticks is 5 cm shorter than the other stick.
What is the length of each stick?

3 A square has an area of 729 cm².

Find the length of one of the sides.

Area
729 cm²

4 Ethel is three times as old as her grandson. Their total age is 108.
How old is Ethel?

5 The length and the width of this rectangle add up to 22 cm.
The area of the rectangle is 112 cm².

Find the length and the width of the rectangle.

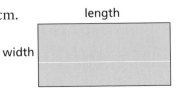

length

width

6 The volume of this cube is 5832 cm³.
Find the length of an edge.

128

Solving equations

Section A

The scales balance in these pictures.
Find the weight of each object.

1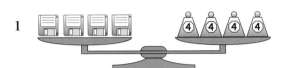

2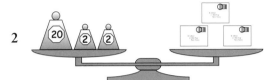

3

4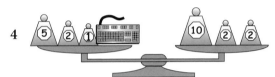

5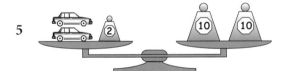

6

7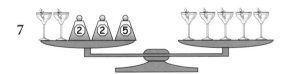

8

9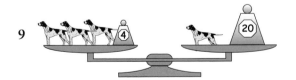

10

Sections B, C and D

1 Here is a set of equations.

(a) $$4x + 3 = 19$$

(b) $$3x + 4 = 19$$

(c) $$3x + 2 = 4x$$

(d) $$3x + 2 = x + 4$$

For each equation

- write down the balance puzzle below that matches it
- solve the puzzle and write down the solution in the form '$x = ...$'

A 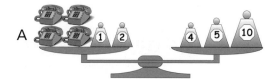 B

C  D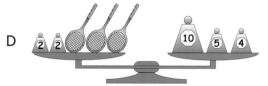

2 Use balancing to solve these equations.
(Show your working clearly and check each answer.)

(a) $2x + 3 = 13$ (b) $5x + 6 = 21$ (c) $4x + 1 = 25$

(d) $7x = 2x + 20$ (e) $3x + 12 = 4x$ (f) $x + 14 = 8x$

3 (a) Write down an equation for this puzzle.
Use x to stand for the weight of a tin.

(b) Solve the equation to find the weight of a tin.

4 Use balancing to solve these equations.

(a) $2x + 1 = x + 5$ (b) $4x + 3 = 3x + 6$ (c) $5x + 2 = x + 10$

(d) $4x + 3 = 2x + 13$ (e) $3x + 6 = 7x + 2$ (f) $2x + 10 = 5x + 1$

5 What does x stand for in each diagram?

(a)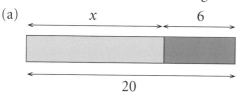

(b)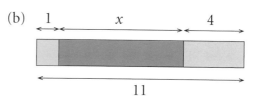

(c)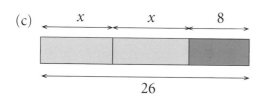

(d)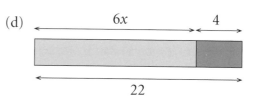

(e)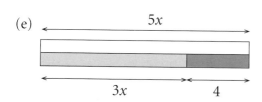

(f)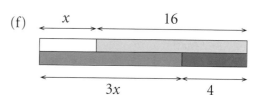

(g)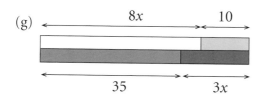

(h)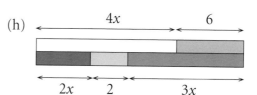

Section E

1 Solve these equations.

(a) $x - 2 = 3$

(b) $x - 8 = 4$

(c) $2x - 2 = 4$

(d) $3x - 2 = 10$

(e) $5x - 1 = 4$

(f) $4x - 1 = 19$

(g) $2x - 1 = x + 4$

(h) $3x - 2 = x + 6$

(i) $2x + 4 = 5x - 2$

(j) $x + 5 = 5x - 3$

(k) $4x - 6 = 2x + 8$

(l) $3x + 5 = 7x - 11$

2 Work out the value of x for each strip.

(a)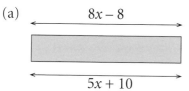

(b)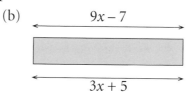

*3 Solve these equations.

(a) $x = 2x - 5$

(b) $4x - 9 = 3x$

(c) $2x = 6x - 8$

(d) $3x - 2 = 2x - 1$

(e) $5x - 10 = 3x - 2$

(f) $6x - 16 = x - 1$

131

Section F

1 (a) Find an expression, in terms of x, for the perimeter of this rectangle. Give your answer in its simplest form.

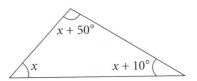

(b) The perimeter of the rectangle is 48 cm. Write down an equation and solve it to find the value of x.

2 (a) Write an expression for the sum of the angles marked in this triangle. Give your answer in its simplest form.

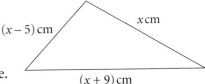

(b) The angles of a triangle add up to 180°. Write down an equation in x and use it to find the value of x.

3 The perimeter of this triangle is 49 cm.

(a) Write down an equation involving x.

(b) Solve your equation to find the value of x.

(c) Write down the lengths of the sides of the triangle.

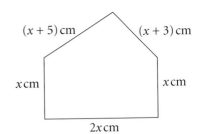

4 (a) Find an expression, in terms of x, for the perimeter of this pentagon. Give your answer in its simplest form.

(b) The perimeter of the pentagon is 62 cm. Write an equation and solve it to find the value of x.

Section G

1 Solve these equations and write each answer as a decimal.
  (a) $4y = 6$
  (b) $10p + 3 = 4$
  (c) $5n - 3 = 1$
  (d) $3x + 2 = x + 5$
  (e) $7y - 1 = 3y + 9$
  (f) $a + 10 = 6a + 7$

2 Solve these equations. Each solution is a negative number.
  (a) $b + 6 = 4$
  (b) $y + 5 = 1$
  (c) $2a + 5 = 1$
  (d) $p + 6 = 3p + 8$
  (e) $5s + 6 = 3s + 2$
  (f) $3n + 2 = 2n - 1$

3 Solve these equations.
  (a) $3t + 5 = 2t$
  (b) $2y - 2 = 5y + 1$
  (c) $m + 2 = 3m - 7$
  (d) $4x + 1 = x - 5$
  (e) $n + 6 = 3n - 5$
  (f) $5p + 2 = 3p - 6$

Mixed questions 5

1 Copy this diagram on to squared paper.

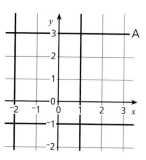

  (a) Label each of the bold lines with its equation.

  (b) On the same diagram, draw and label the line $y = 2x$.

  (c) What are the coordinates of the point where the line $y = 2x$ crosses line A?

2 In May 2002, £1 was worth 1.41 US dollars (US$). How much (to 2 d.p.) was

  (a) £125.75 worth in US$ (b) US$350 worth in £

3 Work out a rough estimate for each of these. Show your working.

  (a) 48.95×5.34 (b) $49.3 \div 1.973$ (c) $\dfrac{97 \times 0.11}{1.99}$ (d) $\dfrac{0.097 \times 3977}{19.87}$

4 Calculate each of the lettered angles. Explain how you know using one of the reasons in the boxes.

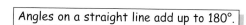

> Angles on a straight line add up to 180°.

> It is an alternate angle.

> Angles in a triangle add up to 180°.

> It is an opposite angle.

> It is a corresponding angle.

5 Shape ABCDE is a regular pentagon. Two sides have been extended to meet at O. Work out the three angles marked.

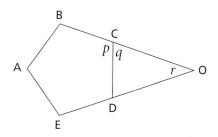

6 Find the area of each of these shapes.

(a)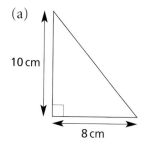

(b)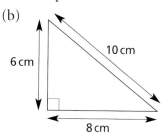

(c)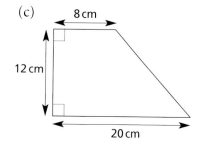

133

7   Cathy's paintings are all 4 cm taller than they are wide.

    She wants to make a painting with an area of 1000 cm².

Width of painting	Height (cm) (= width + 4)	Area (cm²) (target 1000)	Result too small	too large
20	24	480	✓	

Use a table like the one above to find the width of her painting correct to one decimal place.

8   Jo has a normal dice and a coin.
    One side of the coin has a 1 written on it. The other side has a 2 on it.

    Jo rolls the dice and flips the coin.
    She adds the two numbers together to get the score.

    (a) Copy and complete this table to show all the possible scores.

    (b) What is the probability that the score is 5?

    (c) What is the probability that the score is 4 or less?

Score on dice

Score on coin	1	2	3	4	5	6
1			4			
2	3					

9   Write these as ratios in their simplest form.
    (a) A litter has 8 male puppies and 4 female puppies in it.
    (b) A Spanish class has 6 men and 9 women.
    (c) A youth choir has 36 girls and 8 boys.
    (d) A company has 10 women programmers and 25 men.

10 Salvador makes Sangria from red wine, lemonade and orange juice in the ratio $3:2:1$.
    (a) Salvador makes Sangria using 4 litres of lemonade.
        (i)   How much red wine will he need?   (ii) How much orange juice will he need?
    (b) Salvador wants to make 18 litres of Sangria!
         How much of each ingredient will he use?

11 Solve these equations.
    (a) $a + 3 = 12$ (b) $3b = 18$ (c) $c - 1 = 10$
    (d) $d + 12 = 4d$ (e) $e + 12 = 2e + 4$ (f) $8f + 3 = 2f + 15$

12 Solve
    (a) $2n - 5 = 1$ (b) $3n - 1 = n + 5$ (c) $4n + 1 = 5n - 4$

13 (a) Write down and simplify an expression for the perimeter of this rectangular field.
    (b) The perimeter of the field is 64 m. Write down an equation in x and solve it.
    (c) Write down the length of each side of the field and thus work out its area.

$x + 2$ metres

x metres x metres

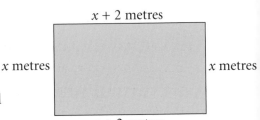

$x + 2$ metres

43 Representing data

Sections A and B

1 This table shows the most popular girls' names for babies in the last century.

Order	1904	1924	1954	1984	1994	1998
1	Mary	Margaret	Susan	Sarah	Rebecca	Chloe
2	Florence	Mary	Linda	Laura	Lauren	Emily
3	Doris	Joan	Christine	Gemma	Jessica	Megan
4	Edith	Joyce	Margaret	Emma	Charlotte	Jessica
5	Dorothy	Dorothy	Janet	Rebecca	Hannah	Sophie

© Crown Copyright 2001 Source: National Statistics

(a) Which name was most popular in 1924?

(b) Which name was the third most popular in 1954?

(c) Which names appear twice in the lists?

2 This train timetable shows times of trains from Three Bridges to Clapham.

Three Bridges	0647	0712	0730	0733	0742	0803	0812	0830
Clapham	0719	0751	0803	0810	0822	0838	0851	0903

(a) Rob arrives at Three Bridges at a quarter past seven. Which train will he catch to Clapham and how long will the journey take?

(b) How many trains depart from Three Bridges for Clapham between 7 a.m. and 8 a.m.?

(c) Which is the slowest train from Three Bridges to Clapham?

3 This table shows results from a class survey.
Some of the numbers have been left out.

	Play a musical instrument	Do not play a musical instrument	Totals
Boys	7	....	16
Girls	......	5	......
Totals	16	......	30

(a) Copy and complete this table.

(b) How many boys are there in the class?

(c) How many girls are there in the class?

(d) What fraction of the girls play a musical instrument?

(e) What fraction of the class play a musical instrument?

Sections C and D

1 Sean sorts through all the music
CDs in his house.

Type of music	Rock	Metal	Classical	Jazz
Number of CDs	30	25	50	12

    (a) Draw a pictogram for the CDs in Sean's house.
Use the symbol ⊙ to represent 10 CDs.

    (b) Draw a bar chart for the CDs in Sean's house.

2 Aaron has carried out a survey on the number of
bedrooms in houses in his area.
His results are shown in this bar chart.

    (a) How many houses had 2 bedrooms?

    (b) How many houses did Aaron use in his survey?

    (c) What was the modal number of bedrooms
in Aaron's survey?

    (d) Aaron's cousin, Jason, does the same survey
where he lives. Here are Jason's results.

Bedrooms	1	2	3	4
Frequency	3	6	9	2

    Draw a bar chart to show Jason's results.

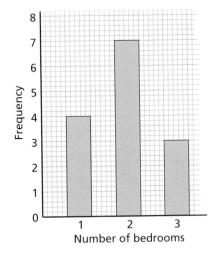

3 This bar chart shows
what some children
in a secondary school
do with their
pocket money.

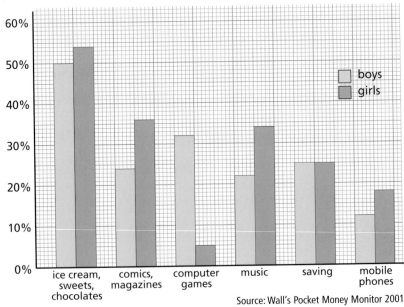

Source: Wall's Pocket Money Monitor 2001

    (a) What do most boys and girls spend their money on?

    (b) For which item is there the biggest difference between boys and girls?

    (c) Are mobile phones more popular with the girls or the boys at this school?

    (d) What percentage of students said that they saved some of their money?

Section E

1 This time series graph shows the average amount of pocket money received by 11- to 13-year-old children between 1991 and 2001.

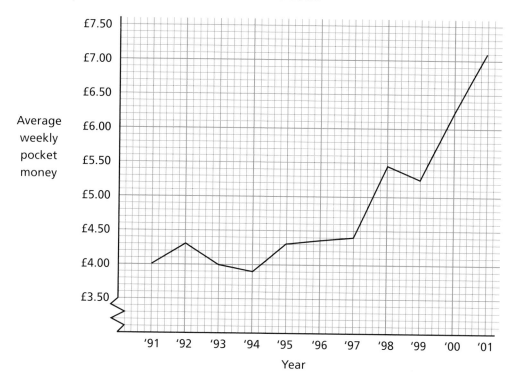

(a) What was the average weekly amount of money given to 11- to 13-year-olds in
   (i) 1993   (ii) 1995   (iii) 2001

(b) In which years did the amount of money received go down?

(c) In which years did the amount of money received rise steeply?

2 This time series graph shows the cost of telephone bills for a household over four years.

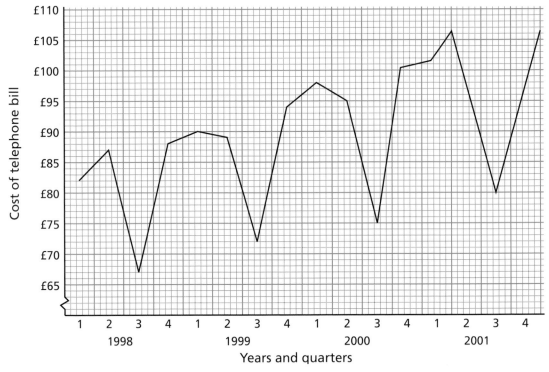

Years and quarters

Quarter 1 is months January, February, March.
Quarter 2 is months April, May, June.
... and so on.

(a) What are the quarterly bills for each of the four quarters in 1998?

(b) From the graph, which quarter is usually the most expensive in a year? Which months are in this quarter?

(c) From the graph, which quarter is usually the least expensive in a year? Which months are in this quarter?

 Explain why you think this is the cheapest quarter.

(d) Between 1998 and 2001 have the bills generally increased, got lower or stayed the same?

44 Calculating with negative numbers

Section A

1 Work these out.

   (a) 5 − 9   (b) ⁻3 + 4   (c) ⁻6 − 1   (d) 0 − 8

   (e) ⁻9 + 6   (f) ⁻4 + 4   (g) 2 − 20   (h) ⁻6 + 20

2 Work these out.

   (a) 4 + ⁻3   (b) 3 + ⁻7   (c) ⁻1 + ⁻4   (d) ⁻10 + ⁻3

   (e) 6 + ⁻4   (f) ⁻5 + ⁻5   (g) 8 + ⁻8   (h) ⁻3 + ⁻4 + ⁻5

   (i) ⁻10 + 4 + ⁻2

3 Find the missing number in each calculation.

   (a) 8 + **?** = 5   (b) 7 + **?** = ⁻1   (c) **?** + ⁻3 = ⁻9   (d) **?** + ⁻4 = 3

4 In a magic square the numbers in each row, each column and each diagonal add to the same total.

   Copy and complete these magic squares.

   (a)

⁻3		8
	2	
		7

   (b)

	3	⁻5
	⁻2	
1		

   (c)

⁻4	1	
3		
⁻2		

5 Work these out.

   (a) 6 − ⁻2   (b) 4 − ⁻7   (c) ⁻2 − ⁻1   (d) ⁻5 − ⁻8

   (e) ⁻1 − ⁻9   (f) 7 − ⁻3   (g) ⁻4 − ⁻4   (h) 9 − ⁻9

6 Work these out.

   (a) 8 + ⁻3   (b) ⁻2 − ⁻5   (c) ⁻2 + ⁻5   (d) 0 + ⁻7

   (e) 0 − ⁻7   (f) ⁻10 + ⁻2   (g) ⁻9 − ⁻3   (h) ⁻3 − ⁻9

7 Choose pairs of numbers from the loop to make these calculations correct.

   (a) ■ − ■ = ⁻4   (b) ■ − ■ = ⁻5

   (c) ■ + ■ = 1   (d) ■ − ■ = 1

                                     1   5   ⁻3   ⁻4

8 Find all the different answers you can get using ⁻4, 6 and + or −.

9 Find all the different answers you can get using ⁻3, ⁻5 and + or −.

139

10 Work out each calculation, use the code and rearrange the letters to find three vegetables.

$^-4$	$^-3$	$^-2$	$^-1$	0	1	2	3
A	B	C	E	O	P	R	T

(a) $\boxed{^-1 + 4}$ $\boxed{5 - 5}$ $\boxed{^-3 + 3}$ $\boxed{^-1 - 3}$ $\boxed{^-5 + 6}$ $\boxed{^-4 + 7}$

(b) $\boxed{1 - ^-1}$ $\boxed{^-2 - ^-2}$ $\boxed{^-1 - ^-4}$ $\boxed{^-2 - 2}$ $\boxed{^-3 - ^-5}$ $\boxed{4 + ^-6}$

(c) $\boxed{8 - 9}$ $\boxed{^-1 + 4}$ $\boxed{^-3 - 3}$ $\boxed{^-8 - ^-5}$ $\boxed{2 + ^-3}$ $\boxed{^-2 + 5}$ $\boxed{^-7 + 9}$ $\boxed{^-1 - ^-1}$

Sections B and C

1 Work these out.

(a) $5 \times ^-3$ (b) $^-4 \times 2$ (c) $^-6 \times 3$ (d) $^-2 \times ^-3$

(e) $6 \times ^-1$ (f) $^-2 \times ^-2$ (g) $3 \times ^-4 \times 2$ (h) $5 \times ^-2 \times ^-6$

2 Copy and complete these multiplication grids.

×	$^-1$	$^-3$	4
3	$^-3$		
$^-2$			
$^-5$			

×	$^-4$	6	8
2			
$^-3$			
$^-10$			

×			
2	$^-2$		
$^-3$		$^-6$	12
		$^-8$	

3 Find the missing number in each calculation.

(a) $8 \times \mathbf{?} = ^-24$ (b) $^-5 \times \mathbf{?} = 35$ (c) $\mathbf{?} \times ^-6 = 36$ (d) $^-4 \times \mathbf{?} = 32$

4 Work these out.

(a) $(^-5)^2$ (b) $(^-3)^2$ (c) $(^-10)^2$

5 Copy and complete these multiplication walls.

(a) 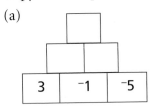   (b) 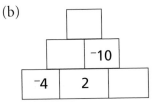   (c)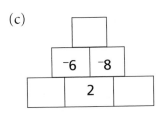

6 Work these out.

(a) $10 \div ^-2$ (b) $^-20 \div 4$ (c) $^-18 \div ^-9$ (d) $^-40 \div ^-5$ (e) $42 \div ^-6$

7 Calculate these.

(a) $5 \times {}^-9$ (b) ${}^-4 \times {}^-6$ (c) ${}^-28 \div 7$ (d) ${}^-64 \div {}^-8$ (e) ${}^-4 \times {}^-10$

(f) ${}^-50 \div 10$ (g) ${}^-8 \times 2$ (h) $36 \div {}^-9$ (i) $6 \div {}^-6$ (j) ${}^-9 \div {}^-9$

8 Find the missing number in each calculation.

(a) ${}^-4 \times \mathbf{?} = 12$ (b) ${}^-70 \div \mathbf{?} = {}^-10$ (c) ${}^-8 \times \mathbf{?} = 40$ (d) ${}^-6 \times \mathbf{?} = 6$

(e) $\mathbf{?} \div {}^-7 = {}^-8$ (f) $90 \div \mathbf{?} = {}^-9$ (g) $\mathbf{?} \times {}^-4 = 16$ (h) ${}^-48 \div \mathbf{?} = {}^-8$

9 Choose pairs of numbers from the loop to make these calculations correct.

(a) $\blacksquare \div \blacksquare = {}^-2$ (b) $\blacksquare \times \blacksquare = {}^-12$

(c) $\blacksquare \div \blacksquare = 4$ (d) $\blacksquare \times \blacksquare = 24$

3    ${}^-4$ ${}^-6$ ${}^-24$

10 Use two numbers from $3, 5, {}^-5, {}^-20$ and either $\times$ or $\div$ to get the following results.

(a) ${}^-15$ (b) 4 (c) 100 (d) ${}^-25$ (e) ${}^-1$ (f) ${}^-60$

Section D

1 Work these out.

(a) $4 + {}^-9$ (b) $3 \times {}^-7$ (c) $3 - 10$ (d) $20 \div {}^-5$ (e) ${}^-3 \times {}^-6$

(f) $5 - {}^-1$ (g) ${}^-7 \times {}^-1$ (h) ${}^-2 - {}^-8$ (i) ${}^-24 \div 6$ (j) ${}^-30 \div {}^-5$

2 Work these out.

(a) $5 + {}^-2 + {}^-1$ (b) $5 - 4 - 3$ (c) ${}^-5 \times 3 \times {}^-2$ (d) ${}^-1 \times {}^-2 \times {}^-3$

3 Calculate these.

(a) $(4 - 7) \times {}^-3$ (b) $2 \times ({}^-3 - 4)$ (c) $({}^-2 - {}^-3) \times 5$ (d) $(2 - 6)^2$

(e) $\dfrac{1 - 7}{3}$ (f) $\dfrac{{}^-6 \times 4}{8}$ (g) $\dfrac{{}^-9 + 5}{{}^-2}$ (h) $\dfrac{({}^-10)^2}{4}$

4 Find the missing number in each calculation.

(a) $9 + \mathbf{?} = 2$ (b) $4 \times \mathbf{?} = {}^-36$ (c) $\mathbf{?} \div 7 = {}^-3$

(d) ${}^-4 - \mathbf{?} = 1$ (e) $8 + {}^-3 + \mathbf{?} = {}^-1$ (f) ${}^-45 \div \mathbf{?} = 5$

5 Simon recorded the lowest temperatures each day for a week in January.

Day	Mon	Tues	Wed	Thurs	Fri	Sat	Sun
Temp °C	${}^-2$	1	${}^-3$	${}^-4$	${}^-2$	${}^-3$	${}^-1$

(a) Which day had the lowest temperature?

(b) What is the difference between the highest and lowest temperatures in the table.

(c) Calculate the mean of the temperatures in the table.

6 (a) The rule to continue this sequence is **– 5**.
   Find the next two terms of the sequence.

   8   3   ⁻2   ....

(b) The rule to continue this sequence is **+ 7**.
   Find the next two terms of the sequence.

   ⁻27   ⁻20   ⁻13   ....

(c) The rule to continue this sequence is **multiply by ⁻2.**
   Find the next two terms of the sequence.

   5   ⁻10   20   ....

7 Here is a number machine chain.

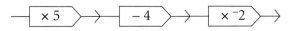

Find the output for each of these inputs.

(a) 3   (b) ⁻1   (c) 5

8 Work these out, use this code and rearrange the letters to find four sports.

⁻4	⁻3	⁻2	⁻1	0	1	2	3	4
A	N	R	I	E	S	K	T	G

(a) ⁻3 + 2   1 × ⁻3   ⁻1 × ⁻2   ⁻3 + 2   7 – 3   ⁻1 + 2

(b) ⁻9 ÷ ⁻3   ⁻2 – ⁻2   1 – 5   ⁻6 ÷ 3   ⁻2 – 2   ⁻8 – ⁻10

(c) ⁻1 × ⁻4   2 – 5   ⁻2 – ⁻1   ⁻1 + 4   ⁻2 × 2   ⁻4 ÷ 2   5 – 3

(d) ⁻3 × ⁻1   ⁻2 ÷ ⁻2   ⁻5 – ⁻1   3 ÷ ⁻3   ⁻6 + 3   ⁻6 ÷ ⁻3   (⁻2)²

Section E

1 Work out the value of each expression when $a = {}^-8$ and $b = 3$.

(a) $a + b$ (b) $a - b$ (c) $b - a$ (d) $2a$ (e) ab

2 What is the value of each expression when $c = 4$ and $d = {}^-5$?

(a) $c - 10$ (b) $2c + d$ (c) $2d + 1$ (d) $c - d$ (e) $2cd$

3 For each set of cards, work out the value when $x = {}^-2$ and $y = {}^-3$.
Put the cards in order of size, smallest to largest, to spell a word.

(a)

$x + y$	$x - y$	$y - x$
A	E	T

(b)

$2x - y$	$2x + 1$	xy	$2y + 1$
A	E	M	T

(c)

$5 + x$	$2 - y$	x^2	$3y$
C	D	E	I

4 What is the value of each expression when $r = {}^-2$, $s = 4$ and $t = {}^-5$?

(a) $r + s + t$ (b) $r - s$ (c) $2s + 2t$ (d) t^2 (e) $r + 2s$

5 Find the value of each of these expressions if $p = {}^-3$ and $q = 6$.

(a) pq (b) $\dfrac{q}{p}$ (c) $p^2 + 1$ (d) $\dfrac{pq}{2}$ (e) $\dfrac{p + q}{3}$

6 Use the formula $v = u - 10t$ to find the value of v if $u = 5$ and $t = 2$.

7 Use the formula $C = 4p + q$ to find the value of C when $p = 2$ and $q = {}^-5$.

8 Work out the values of each expression below when $w = {}^-2$, $x = 3$ and $y = {}^-3$.
Use the code below to change the values to letters. Each line is a different word.
Unscramble each word to read a message.

$w + 2x$ $x - 5$ wx $x + y$ $2x + y$

$y - x$ $w + x$ $y - 1$

$2w + 1$ $w + y$

$2y$

$x - y$ $x - w$ wy $\dfrac{x}{y}$ $x + 2y$ $y + 5$

$^-6$	$^-5$	$^-4$	$^-3$	$^-2$	$^-1$	0	1	2	3	4	5	6
A	B	C	E	H	L	M	N	P	S	T	U	Z

45 Metric units

Section A

1 For each of the following statements decide whether or not
the measurements are sensible. If the statement is not sensible,
rewrite it with a reasonable estimate for the measurement.

(a) John has to walk 2 km to school each day.

(b) I bought 5 g of potatoes at the greengrocers.

(c) A can of cola holds 330 litres of liquid.

(d) My teacher is 20 m tall.

(e) A box of chocolates weighs 500 g.

(f) The capacity of the petrol tank in a car is 5 litres.

2 Write down the metric units you would use to measure the following.

(a) the width of this book

(b) the distance from Glasgow to London

(c) the weight of a sack of potatoes

(d) the amount of liquid in a full cup of tea

3 Change these lengths into centimetres.

(a) 7 m   (b) 60 mm   (c) 2.7 m

(d) 350 mm   (e) 0.8 m   (f) 3 mm

4 Change these lengths into metres.

(a) 3 km   (b) 250 cm   (c) 28 cm

(d) 5.2 km   (e) 0.7 km   (f) 9 cm

5 Change these weights into grams.

(a) 5 kg   (b) 0.7 kg   (c) 15 kg   (d) $\frac{3}{4}$ kg

6 Change these weights into kilograms.

(a) 4000 g   (b) 3400 g   (c) 200 g   (d) 355 g

7 Change these measurements.

(a) 3 litres into ml   (b) 4.5 litres into ml   (c) 0.8 litres into ml

(d) 3000 ml into litres   (e) 1230 ml into litres   (f) 650 ml into litres

Sections B and C

1 Approximately how many litres are there in
   (a) 4 pints   (b) 18 pints   (c) 30 pints   (d) 9 pints

2 Approximately how many pints are there in
   (a) 4 litres   (b) 18 litres   (c) $8\frac{1}{2}$ litres   (d) $25\frac{1}{2}$ litres

3 Approximately how many millilitres are there in
   (a) 4 pints   (b) $\frac{1}{5}$ pint   (c) 0.75 pint   (d) $\frac{1}{10}$ pint

4 One day Derek the milkman delivered the following amounts of each type of milk.

   Gold top $4\frac{1}{2}$ gallons
   Silver top   18 gallons
   Semi-skimmed   16 gallons
   Skimmed   9 gallons

   (a) How many pints of each type of milk were delivered?

   (b) Approximately how many litres of each type of milk were delivered?

   (c) Change the amounts in gallons to litres accurately by multiplying by 4.5.

5 Rewrite these lengths using the metric measurements
   indicated in the brackets.

   (a) 8 feet (cm)   (b) 4 feet (cm)
   (c) $3\frac{1}{2}$ feet (cm)   (d) 45 feet (m)
   (e) 240 feet (m)   (f) 9000 feet (m)

Metric	Imperial
30 cm	1 foot
1 metre	3 feet
1 kg	2 lb
8 km	5 miles
4.5 litres	1 gallon

6 Rewrite the following statement using the metric
   measurements indicated in the brackets.

   (a) I travelled 400 miles to get to Scotland. (km)

   (b) The petrol tank in my car holds 15 gallons. (litres)

   (c) The weight of all the luggage we took on holiday
        was 100 lb. (kg)

   (d) My rubber plant has grown to a height of $7\frac{1}{2}$ feet. (m)

46 *Finding and using formulas*

Section A

1 Copy this crossnumber grid.
Complete it using the clues below.
For the clues $x = 7$, $y = 9$ and $z = 5$.

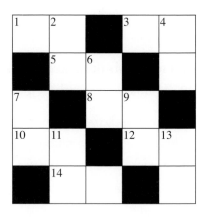

CLUES ACROSS
  1 $x + y + z$
  3 $y + 2$
  5 $3y$
  8 $2x + 2z$
 10 $10 + \frac{y}{3}$
 12 $10z - 2$
 14 $4y + 4z$

CLUES DOWN
  2 $x + z$
  4 $2y - 1$
  6 $8y$
  7 $8z - 9$
  9 $10z - 6$
 11 $7z$
 13 $12z + 20$

2 Work out each of these.

  (a) $2(x + 5)$ when $x = 3$ (b) $3(x + 4)$ when $x = 7$ (c) $5(x - 1)$ when $x = 4$

  (d) $2(x - 3)$ when $x = 9$ (e) $4(x + 2)$ when $x = 8$ (f) $4x + 8$ when $x = 8$

3 Work out each of these when $x = 10$.

  (a) $3x + 1$ (b) $\frac{x}{2} + 4$ (c) $\frac{x + 4}{2}$

  (d) $\frac{1}{2}x + 5$ (e) $2x + 1$ (f) $2(x + 1)$

  (g) $3(x - 5)$ (h) $3x - 5$ (i) $2(20 + x)$

Sections B and C

1 If $x = 3$, $y = 10$ and $z = 6$
work out the value of

  (a) $10 - 2x$ (b) $10 - y$ (c) $30 - 2z$

  (d) $30 + \frac{y}{2}$ (e) $30 - \frac{z}{2}$ (f) $20 - \frac{z}{2}$

2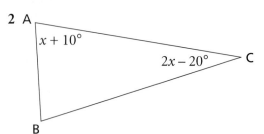

If $x = 30°$

  (a)   work out the size of angle BAC

  (b)   work out the size of angle ACB

  (c)   what is the size of angle ABC

3 Find expressions for the areas of these shapes.

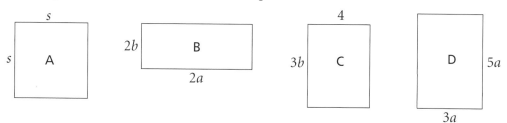

4 Find four pairs of matching algebra cards.

Which card is the odd one out?

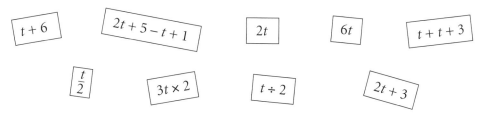

Check that your pairs are correct by trying $t = 10$ in each card.

5 If $x = 4$, $y = 3$ and $z = 10$ work out the value of

(a) xy (b) $2yz$ (c) $2xy$ (d) $10xz$

Sections D and E

1 Work out these when $x = 12$, $y = 20$ and $z = 2$.

(a) xz (b) $5 + yz$ (c) $3yz$ (d) $30 - 3z$ (e) $\dfrac{y}{z}$

2 Work out these when $x = 3$, $y = 5$ and $z = 1$.

(a) $4(x + z)$ (b) $2(y + x)$ (c) $3(10 - y)$ (d) $2(y - z)$

3 Work out these when $x = 3$, $y = 10$ and $z = 1$.

(a) x^2 (b) $2x^2$ (c) y^2 (d) $y^2 - 8$ (e) $3x^2$

4 Find four pairs of matching algebra cards. Which card is the odd one out?

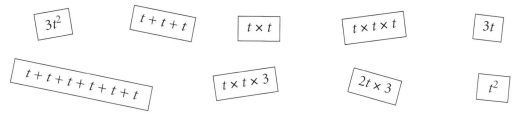

5 For each of these arrow diagrams
    (i) work out the result when 10 is put in
    (ii) If n is put in, write an expression for the result in terms of n.

(a) ◯ —×2→ ◯ —+3→ ◯   (b) ◯ —+2→ ◯ —×3→ ◯

(c) ◯ —÷2→ ◯ —+4→ ◯   (d) ◯ —+2→ ◯ —÷4→ ◯

Section F

1 (a) A rope is 7 metres long. If 2 metres is cut off it, how much is left?

  (b) A rope is 10 metres long. If p metres is cut off it, how much is left?

2 Pencils cost n pence each. Biros cost 8 pence more.

  (a) Write an expression, in terms of n, for the cost of 8 pencils.

  (b) Write an expression, in terms of n, for the cost of 1 biro.

  (c) Write an expression, in terms of n, for the cost of 3 biros.

3 To find the cost of mending burst pipes, a plumber uses the formula

$$C = 50 + 20h$$

where C is the cost in pounds and h is the number of hours it takes.

Calculate the cost of mending a pipe which takes

  (a) 2 hours   (b) 4 hours   (c) $\frac{1}{2}$ hour

4 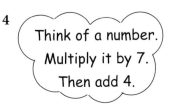 (a) If Maria starts with 5, what is her result?

    (b) If Sam starts with n, what is his result?

Think of a number.
Multiply it by 7.
Then add 4.

5 This pattern of matches has the formula $n = 3s + 1$ where n is the number of matches and s is the number of squares.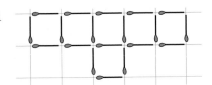

  (a) How many matches are needed for a pattern with 10 squares?

  (b) If a pattern uses 37 matches, how many squares does it have?

Section G

1 This graph converts kilograms (kg) to pounds (lb).

(a) How many pounds are there in 3 kg?

(b) If I buy 4 lb of apples, what is the weight in kg?

(c) How many kilograms of potatoes are in a 7 lb bag?

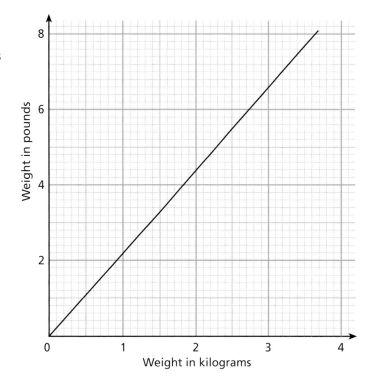

2 Dale wants to join a gym.
There is a £60 joining fee and each session attended costs £2.
A formula for the total cost is $C = 60 + 2n$
where C is the total cost in £, and n is the number of sessions attended.

(a) Copy and complete this table.

Sessions (n)	5	10	15	20	25	30
Total cost in £ (C)	70					

(b) Draw and label axes like the ones on the right.
Plot the points from your table and join them with a line.

(c) Use your graph to answer these questions.

(i) How much will Dale pay if he attends 12 sessions?

(ii) How much will he pay if he attends 28 sessions?

(iii) If he pays £98, how many sessions has he attended?

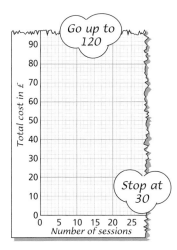

47 *Working with percentages*

Section B

1 Work out
   - (a) 25% of £24
   - (b) 50% of 32p
   - (c) 75% of 12 litres
   - (d) 50% of £35
   - (e) 25% of 22 m
   - (f) 75% of 10 kg

2 In a sale, prices are reduced by 25%.
   How much will be taken off the price of the following items?
   - (a) a television costing £300
   - (b) a mini hi-fi costing £1400
   - (c) a computer costing £2400

3 Work out
   - (a) 10% of £80
   - (b) 30% of £80
   - (c) 70% of £80
   - (d) 40% of £80

4 Work out
   - (a) 20% of £70
   - (b) 30% of £40
   - (c) 60% of 90 g
   - (d) 80% of 110 g

5 How much extra will have to be paid on the following?
   - (a) a 50p bus fare increased by 20%
   - (b) a £30 bottle of perfume increased by 30%
   - (c) a television set costing £350 increased by 40%

6 Work out
   - (a) 10% of £320
   - (b) 5% of 160 g
   - (c) 15% of 360 ml

7 Work out
   - (a) 15% of 30 kg
   - (b) 45% of 400 g
   - (c) 55% of £80

8 A packet of biscuits are described as 65% fat free.
   - (a) What percentage of the biscuits are fat?
   - (b) If the packet weighs 500 g, how much fat do the biscuits contain?

9 In a school, 85% of teachers are female.
   - (a) What percentage of the teachers are male?
   - (b) If there are 80 teachers in the school, how many are female?

Section C

1 Calculate
   (a) 36% of 75
   (b) 16% of £90
   (c) 42% of 360
   (d) 26% of £65
   (e) 38% of £23
   (f) 93% of 48 kg

2 Alison always saves 6% of her monthly wages. How much will she save if she earns
   (a) £1450
   (b) £1350
   (c) £2420

3 13% of a type of Dutch cheese is protein. How much protein is there in a piece of cheese weighing 300 g?

4 A car bought for £12 600 had lost 53% of its value after 3 years.
   (a) How much value had it lost in money?
   (b) What was the value of the car after 3 years?

5 A travel company advertises a discount of 12%.
   What will be the discount on the following holidays?
   (a) a fly-drive holiday for £816
   (b) an activity holiday for £482

6 250 candidates entered for a GCSE maths exam.
   The percentage of students given each grade is shown in the table below.

Grade	A*	A	B	C	D	E	F	G
Percentage	2%	4%	14%	32%	22%	16%	4%	6%

   How many candidates gained each grade?

Section D

1 Write these as percentages.
   (a) 12 out of 25
   (b) 11 out of 20
   (c) 14 out of 50
   (d) 70 out of 200
   (e) 210 out of 300
   (f) 350 out of 1000

2 In a class of 25 students, 6 of them walk to school.
   What percentage of the class walk to school?

3 In a bowl of 20 pieces of fruit there are 9 clementines.
   What percentage are clementines?

4 In a bunch of 10 freesias, 2 are white, 3 are yellow and 5 are lilac.
   What percentage of the bunch of freesias are
   (a) white
   (b) yellow
   (c) lilac
   (d) not yellow

5 In a school of 400 pupils, 32 of them are left-handed.
What percentage are left-handed?

6 For each of the amounts below
- write it as a fraction
- change the fraction to its simplest form
- write the fraction as a percentage

(a) 45 out of 90   (b) 9 out of 60   (c) 21 out of 70

Section E

1 Work out as percentages
(a) 36 out of 450   (b) 54 out of 72   (c) 56 out of 350
(d) 6 out of 15   (e) 168 out of 480   (f) 189 out of 450

2 I collected 75 apples from the tree in my garden. 36 of them were bad.
(a) What percentage were bad?   (b) What percentage were edible?

3 In his end-of-term tests Matthew received the following results.

English $\frac{68}{80}$ French $\frac{132}{150}$ Maths $\frac{33}{60}$ Science $\frac{72}{90}$

What percentage did he receive for each subject?

4 Write these as percentages to the nearest 1%.
(a) 33 out of 45   (b) 56 out of 65   (c) 27 out of 70
(d) 19 out of 24   (e) 320 out of 350   (f) 38 out of 52

5 Nicola was given £45 for Christmas.
She writes down how she spends the money.

CDs £19   Books £12   Stationery £6   Make-up £8

Work these out as a percentage of the total amount.
Check that they add up to 100%.

6 Chardonnay carries out a survey on how often students
in her year group read a newspaper.
These are her results.

Every day 12   Sometimes 19   Never 23

Work these out as a percentage of the total number of people in the survey.

48 Coordinates

Section A

1 The diagram shows five polygons plotted on a coordinate grid.

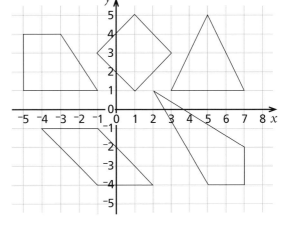

Find each of the following polygons and list its coordinates.

(a) isosceles triangle (3, 1)...

(b) kite

(c) square

(d) parallelogram

(e) trapezium

2 The diagram shows three points A, B and C.

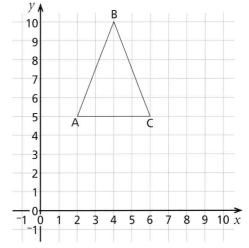

(a) Give the special name for triangle ABC.

(b) Shape ABC has one line of symmetry. Write down the coordinates of four points on the line of symmetry.

(c) What is the equation of the line of symmetry in (b)?

(d) What are the coordinates of point E, where ABCE is a rhombus?

3 Draw a grid on squared paper with the x and y-axes going from ⁻4 to 7.

(a) Mark points A(0, 5), B(2, 3), C(0, 1). Join A to B to C.

(b) Mark a point D so that ABCD is a square.

(c) Mark a point E so that ABCE is a kite.

(d) Draw a line of symmetry for the kite ABCE.

(e) What is the equation of the line of symmetry?

Sections B and C

1 (a) Using the coordinate grid,
write down the coordinates
of the points that are
halfway between

(i) A and B

(ii) B and C

(iii) C and D

(iv) B and E

(b) C is halfway between B and
a new point X.
What are the coordinates
of point X?

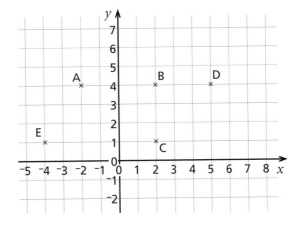

2 On centimetre squared paper, draw a grid with the x and y-axes going from ⁻1 to 7.

(a) Draw the square whose corners are at A(1, 4), B(4, 7), C(7, 4), D(4, 1).

(b) Work out the area of square ABCD.

(c) Mark the midpoints of each side of this square.
Join them to form a new quadrilateral. What type of quadrilateral is this new one?

(d) Find its area.

3 The diagram shows four cubes fixed together.
Point A has coordinates (1, 2, 0).
Write down the coordinates of B, C and D.

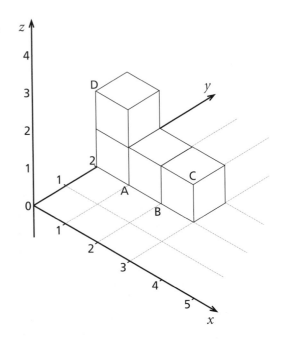

154

49 *Problem solving with a calculator*

Section A

1 Six people have a picnic lunch.
They agree to share the cost equally.

They spend £14.80 on sandwiches, £8.46 on cakes and £4.28 on drinks.

(a) How much do they spend altogether?

(b) How much does each person pay?

2 Four adults and a child go into a theme park.
The total cost of their tickets is £48.95.
A child ticket costs £7.55.

(a) What is the total cost of the four adult tickets?

(b) How much does one adult ticket cost?

3 Jan buys some bottles of cola and a bottle of orange squash.
The bottles of cola cost £1.19 each.
The bottle of orange squash costs £1.49.
The total cost is £9.82.

(a) What is the total cost of the bottles of cola?

(b) How many bottles of cola does Jan buy?

Section B

1 Snappy Film Processing charges £2.55 to develop a film, plus 18p for each print.

(a) How much do they charge altogether to develop a film and make 24 prints?

(b) Gary paid £8.13 to have a film developed and some prints made.
How many prints did he have?

2 Three people make cakes and sell them. They agree to share the profit equally.
The ingredients for making the cakes cost £4.88.
The cakes are sold for £12.35 altogether.

How much profit does each person get?

3 The mileometer on Arnie's car read 45 928 miles at the start of a week.
During that week, Arnie used the car only to drive to and from work.
He drove to work and back four times.
At the end of the week the mileometer read 46 060 miles.

How far is it from Arnie's home to work?

Section C

Use these exchange rates in the questions below.

> £1 = 2.71 Australian dollars
> £1 = 69.55 rupees (India)
> £1 = 183 yen (Japan)

1 Change £45.80 into Australian dollars.
   Give the answer to the nearest dollar.

2 Change 315 Australian dollars to £, to the nearest penny.

3 (a) Marie goes to India. She changes £200 into rupees.
      How many rupees does she get?

   (b) When Marie leaves India she changes 1860 rupees into £.
      How much does she get, to the nearest penny?

4 In Japan, Bharat sees a pocket TV on sale for 14 900 yen.
   How much is this in £, to the nearest penny?

Section D

1 Shop A sells glasses in boxes of six for £2.70 a box.
   The same glasses are on sale in shop B in boxes of eight for £3.80 a box.
   Which is the better deal? Explain how you decide.

2 Shop A sells freezer bags in packs of 40 for £1.40.
   The same bags are sold in shop B in packs of 25 for 80p.
   Which is the better deal? Explain how you decide.

3 Greenfingers Garden Centre sells liquid fertiliser in 4.5 litre cans for £5.85.
   The Ace of Spades garden shop sells the same fertiliser in 2.5 litre cans for £3.10.
   Which is the better deal? Explain how you decide.

4 A 200 ml bottle of Hacker's Cough Mixture costs £1.80.
   A 350 ml bottle costs £2.94.
   Which bottle is better value? Explain your answer.

5 Golden Film Processors charge £2.75 to develop a film, plus 25p for each print.
   Silver Processors charge nothing for developing but charge 35p per print.

   (a) Jack wants a film developed and 24 prints made.
      Which processor is cheaper? Explain.

   (b) Jill wants a film developed and 36 prints made.
      Which processor is cheaper? Explain.

50 Brackets

Section A

1 Work out these expressions when $x = 3$, $y = 5$ and $z = 6$.

   (a) $4x$ (b) xy (c) $2(x + z)$ (d) $3y - z$ (e) $3(y - x)$

   (f) $x + 2z$ (g) $\frac{1}{2}(x + y)$ (h) $z - 2x$ (i) $yz - 3x$ (j) $2(x + y + z)$

2 Simplify each of these expressions.

   (a) $p + p + p + p$ (b) $p + 5 + p - 2$ (c) $q + 8 + q - 3 + q$

   (d) $2 + q - 1 + q - q$ (e) $r + 1 + r + 2 + r + 3$

3 Simplify

   (a) $3a + 5a$ (b) $6b + 4 + b$ (c) $5c + 1 - 2c + 4$

   (d) $3d + 5 - 2d - 1$ (e) $6e + 1 + 2e + 1 - e$ (f) $11 + 4f - 7 - 3f$

4 Simplify

   (a) $3r + 2s + r + 4s$ (b) $4s + 5 + 2t + 3s$

   (c) $6t + 3u - 2t + 7u$ (d) $5 + u + 3v - 2 - v + 4u$

   (e) $3v + 4w - 2v + 5w - v$

5 Find and simplify an expression for the perimeter of each of these shapes.

  (a)   (b)   (c)

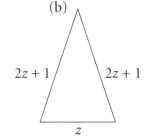

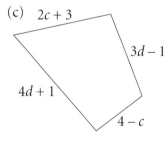

  (d)   (e)

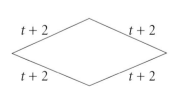

6 Simplify

   (a) $a \times a$ (b) $3b \times b$ (c) $2c \times 5c$ (d) $d \times 7d$ (e) $6e \times 6e$

7 Simplify

  (a) $2f \times g$ (b) $h \times 5j$ (c) $3k \times 7l$ (d) $m \times n$ (e) $7p \times 5q$

Sections B and C

1 There are three pairs of equivalent expressions here.
  Pair them up and find the odd one left over.

  $5(c + 1)$ $5(c + 2)$ $5c + 50$ $5c + 10$ $5(c + 10)$ $5c + 2$ $5c + 5$

2 Multiply out the brackets in each of these expressions.

  (a) $2(a + 3)$ (b) $4(b - 5)$ (c) $8(c - 2)$ (d) $5(4 + d)$ (e) $7(e - 3)$

3 Multiply out the brackets in each of these expressions.

  (a) $5(2a + 3)$ (b) $2(3b - 4)$ (c) $7(2 + 3c)$ (d) $3(4d - 1)$ (e) $9(5e + 7)$

4 Find what is missing in each of these.

  (a) $4(3v - 8) = 12v - ?$ (b) $?(3w + 5) = 6w + 10$ (c) $5(x - ?) = 5x - 20$

5 Find what is missing in each of these.

  (a) $4a + 6 = ?(2a + 3)$ (b) $15b - 12 = ?(5b - 4)$ (c) $40 + 70c = ?(4 + 7c)$

6 Factorise each of these expressions.

  (a) $2p + 12$ (b) $3q + 6$ (c) $6r + 6$ (d) $7s - 14$ (e) $5t + 20$

7 Factorise each of these expressions.

  (a) $4u + 10$ (b) $6v - 9$ (c) $10w + 5$ (d) $15x - 10$ (e) $12y + 2$

8 Factorise each of these as far as you can.

  (a) $8a + 4$ (b) $10b - 25$ (c) $16c - 12$ (d) $20d + 25$ (e) $24e - 32$

9

E	A	C	N	H	R	M	L	I	S
2	3	4	5	6	$x + 1$	$x + 2$	$2x + 1$	$2x + 3$	$3x + 2$

Factorise the expressions below as far as you can.
Use the letters in the table above to find three objects.

  (a) $10x + 15$ $6x + 3$

  (b) $2x + 2$ $3x + 6$ $6x + 12$

  (c) $12x + 6$ $6x + 4$ $8x + 12$

Sections D and E

1 Copy each of these, filling in what is missing.

    (a) $x(x + 5) = x^2 +$ **?**
    (b) $y(y - 7) =$ **?** $- 7y$
    (c) $z(3z - 2) = 3z^2 -$ **?**

2 Multiply out each of these expressions.

    (a) $a(a + 9)$
    (b) $b(b - 6)$
    (c) $c(4 + c)$

    (d) $d(3d + 5)$
    (e) $e(1 - 2e)$
    (f) $f(10f - 9)$

3 There are three pairs of equivalent expressions here, and one left over.
Find the three pairs and multiply out the one left over.

$c(2c + 1)$ $2c^2 + 5c$ $c(3c + 5)$ $2c^2 + c$ $3c^2 + 5c$ $c(c + 1)$ $c(2c + 5)$

4 Factorise each of these expressions as fully as you can.

    (a) $s^2 + 4s$
    (b) $t^2 - 9t$
    (c) $7v + v^2$
    (d) $x^2 + 8x$

    (e) $y^2 + 5y$
    (f) $12z + z^2$
    (g) $2u^2 + 3u$
    (h) $3w - 5w^2$

5 Expand these expressions.

    (a) $3(c + d)$
    (b) $2(x - y)$
    (c) $5(2r + t)$

    (d) $4(3g - 2h)$
    (e) $2(4p - 7q)$
    (f) $6(5m + 4n)$

6 Factorise these expressions.

    (a) $4u + 4v$
    (b) $2a + 8b$
    (c) $5m - 20n$
    (d) $9c + 12d$

7 Simplify the following expressions.

    (a) $2(a + 3) + 4$
    (b) $8 + 5(b - 1)$
    (c) $3(c + 4) + 2c$

    (d) $6(d + 2) - d$
    (e) $7e + 2(4 - e)$
    (f) $5(f - 3) - 2f$

8 Simplify the following expressions.

    (a) $3(2u + 1) + 5u$
    (b) $4 + 5(1 + 3v)$
    (c) $8(2w + 3) - 6w$

    (d) $5(4x + 2) - 10$
    (e) $8y + 2(4 - 3y)$
    (f) $10(5z - 3) - 19z$

9 Simplify

    (a) $3(c + 4) + 5(c - 1)$
    (b) $6(2 + 3d) + 2(d - 4)$
    (c) $5(3e + 1) + 2(7e - 4)$

Sections F and G

1 Simplify the following expressions.

(a) $12 - (a + 1)$ (b) $9 - (b - 2)$ (c) $5c - (2c - 3)$

(d) $9d - (5d + 7)$ (e) $15 - (10 - 4e)$ (f) $8f - (3 - 5f)$

2 Simplify each of these.

(a) $15 - 2(m + 3)$ (b) $9n - 3(n - 2)$ (c) $7p - 5(4 - p)$

(d) $6 - 4(q + 1)$ (e) $15r - 2(2r - 5)$ (f) $8 - 3(4 - s)$

3 Simplify these expressions.

(a) $4g + 5 - 2(g + 1)$ (b) $6(h + 4) - 2(h + 3)$ (c) $4(3 + j) - 2(j - 5)$

(d) $5(k + 7) - 2(k - 3)$ (e) $4(2 + 3l) - 2(l + 5)$ (f) $8(1 - m) - 5(2 - 3m)$

4 There are three pairs of equivalent expressions here, and an odd one out. Which is the odd one out?

$(5x + 1) - 2(x + 1)$

$(5x + 4) - (3x + 1)$ $3(2x + 1) - 2(2x - 1)$ $(4x - 3) - (x - 2)$

$(4x - 3) - 2(x - 1)$ $3(3x + 2) - (7x + 1)$ $(6x + 1) - 2(2x - 1)$

5 Expand each of these expressions.

(a) $4(a + 7)$ (b) $5(b - 6)$ (c) $2(9c + 1)$ (d) $6(3 - 2d)$ (e) $7(5e + 2)$

6 Factorise these as much as you can.

(a) $5r + 20$ (b) $4t - 8$ (c) $9v + 3$ (d) $8w + 6$ (e) $20x - 8$

7 Factorise as fully as possible

(a) $t^2 - 5t$ (b) $k^2 + 10k$ (c) $5k^2 + 8k$ (d) $6c^2 + c$ (e) $3k^2 - 7k$

8 Simplify

(a) $10a \div 5$ (b) $8b^2 \div 2$ (c) $\frac{9x}{3}$ (d) $\frac{15y^2}{3}$ (e) $\frac{18z^2}{6}$

9 Expand and simplify

(a) $4(a + 1) + 3a$ (b) $5(b - 7) + 40$ (c) $7c + 2(c - 3)$

(d) $9 - 3(2 - d)$ (e) $8e - 2(e + 6)$ (f) $12 - (f - 1)$

(g) $4g + 10 - 2(g + 3)$ (h) $5(h + 1) - 2(h - 2)$

10 Expand and simplify

(a) $a + 2(a + b)$ (b) $3(x + y) + 2x - y$ (c) $4p - 5q + 2(p + q)$

51 Navigation

Section B

1 This is a plan of a formal garden, drawn to scale, where 1 cm represents 2 m.

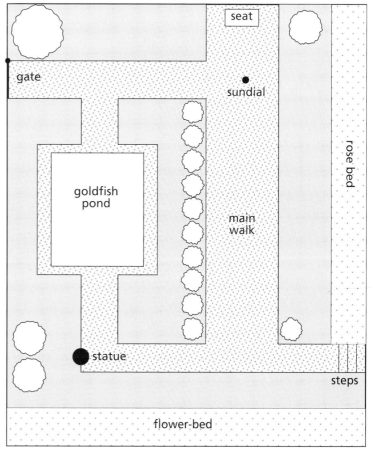

(a) Use the plan to work out what the following measurements would be in real life.

    (i) the width of the main walk

    (ii) the length of the goldfish pond

    (iii) from the gate to the sundial

    (iv) the width of the goldfish pond

(b) A new flower-bed is to be 3 m wide. How wide would it be on the plan?

(c) An edging fence is to be put along the long edge of the rose bed.

    (i) How long will this edging be?

    (ii) The edging is sold in 3 m sections. How many sections will be needed?

    (iii) If sections cost £5.50 each, how much will the total cost of the edging be?

2 This tourist map is drawn to a scale of 1 cm to 5 km.

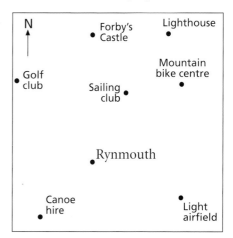

(a) How far is it in real life, in a straight line, from

    (i) The light airfield to Forby's Castle

    (ii) Rynmouth to the lighthouse

    (iii) The mountain bike centre to Light airfield

(b) What do you find 10 km north-east of the sailing club?

(c) How far is the golf club from Rynmouth?

(d) In which direction is the golf club from Rynmouth?

Section D

The map shows mountain peaks at A, B, C, D and E, and two towns, Lowton and Highton.
The scale of the map is 1 cm to 2 km.

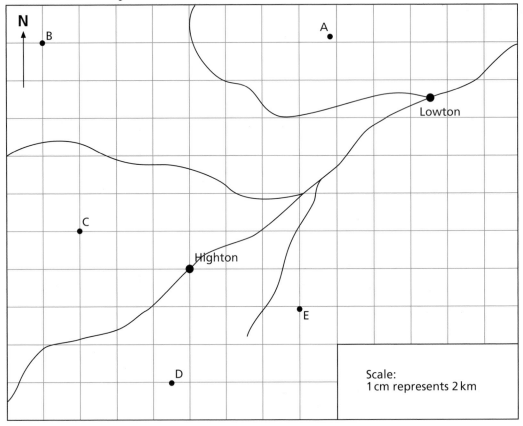

1 How far is it in a straight line from Highton to Lowton in real life?

2 Which mountain peak is on a bearing 110° from Highton?

3 What is the bearing from Highton of mountain peak A?

4 What is the bearing from Highton of mountain peak D?

5 A climber stands on the summit of peak B.
   (a) How far away is peak C?
   (b) How far away is peak A?
   (c) How much further away from the climber is peak A than peak C?
   (d) What is the bearing from peak B of peak A?
   (e) What is the bearing from peak B of peak C?

6 Which peak is on a bearing 225° from Lowton?

7 What is the bearing of peak C from peak E?

52 *Pie charts*

Sections A and B

1 Calculate

(a) $\frac{1}{4}$ of 600   (b) $\frac{1}{5}$ of 300   (c) $\frac{1}{3}$ of 240   (d) $\frac{1}{8}$ of 200

2 Calculate

(a) 25% of 360   (b) 10% of 270   (c) 20% of 360   (d) 40% of 200

3 This pie chart shows the sales of popcorn at a stall one afternoon.

In total there were 300 containers of popcorn sold.

(a) What was the most popular size?

(b) What fraction of sales was the family tub?

(c) How many family tubs were sold?

(d) What fraction of sales was the large carton?

(e) How many large cartons were sold?

(f) Calculate the size of the angle for the standard tub.

(g) How many standard tubs were sold?

(h) How many mega tubs were sold?

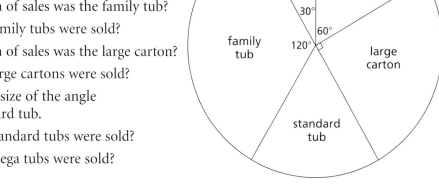

4 Some pupils were asked which type of house they lived in.

The pie chart shows the type of house lived in by these pupils.

(a) What fraction of the pupils lived in a terraced house?

(b) 120 pupils lived in a terraced house. How many pupils were asked altogether?

(c) (i) Measure the angle for semi-detached house.

    (ii) What fraction of the pupils lived in a semi-detached house?

    (iii) How many pupils lived in a semi-detached house?

(d) How many pupils lived in a flat?

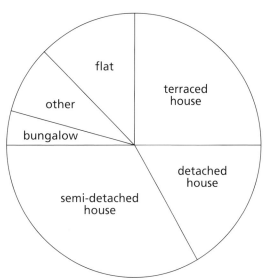

Section C

1 The pie chart shows the main reason why 1800 people visited a library.

(a) How many people does 1° represent?

(b) (i) Measure the angle for CDs and videos.

(ii) How many people came to borrow or return CDs or videos?

(c) How many people came to use the internet?

(d) How many people came to read newspapers?

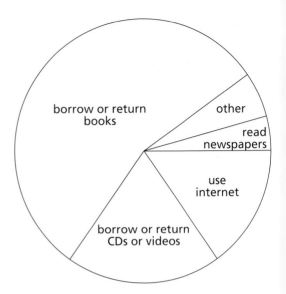

2 In a survey, 216 people were asked which of four types of TV programme they liked best. Their replies are shown in the pie chart.

(a) What type of programme was the mode?

(b) Measure the angle for quiz shows.

(c) How many people liked quiz shows best?

(d) Work out the number of people who liked these best.

(i) soaps (ii) films

3 Some senior citizens were asked about their favourite pets.

The results are shown in the pie chart.

(a) What fraction liked a dog best?

(b) 30 people chose dog. How many people were asked altogether?

(c) How many people chose a cat as their favourite pet?

(d) How many chose a budgie as their favourite pet?

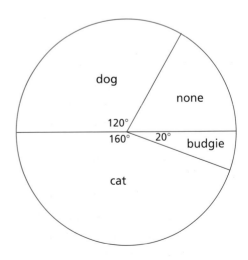

Section D

1 Joe recorded the hair colour of people in his class.

Hair colour	Frequency
Blonde	5
Black	7
Brown	12
Red	2
Other	4

(a) How many people are in his class?

(b) Work out what angle would represent one person in a pie chart.

(c) What will be the angle for people with blonde hair?

(d) Work out all the angles and draw the pie chart. Label each sector.

2 Susan did a survey to find out which type of fiction books people most like to read.

The table shows her results.

Type of book	Frequency
Romance	10
Fantasy	15
Science fiction	7
Crime	4
Other	9

(a) How many people took part in Susan's survey?

(b) Work out what angle would represent one person in a pie chart.

(c) Draw a clearly labelled pie chart to illustrate this information.

3 In a survey, 144 people were asked about their use of computers at home.

The table shows the most popular uses of the computer.

Draw a clearly labelled pie chart to show this information.

Computer use	Frequency
Games	40
Email	36
Word processing	24
Internet shopping	10
Research	12
Other	22
Total	**144**

4 The table shows a summary of an investigation into where in the home 900 accidents occurred.

Where accident occurred	Kitchen	Stairs	Bathroom	Living room	Other
Frequency	400	225	150	90	35

(a) Draw a clearly labelled pie chart to show this information.

(b) In another investigation into accidents in the garden, 400 accidents were investigated. In a pie chart showing the information, the angle for accidents involving lawnmowers was 135°.

How many of the accidents involved lawnmowers?

Section E

1 The table shows the composition of Crunchy Peanut Butter.

Protein	Carbohydrate	Fat	Fibre	Other
25%	10%	51%	7%	7%

Draw and label a pie chart to show this information.

2 Robert asked the pupils in his class which type of chocolate they preferred.

Type of chocolate	Frequency
Milk	11
Plain	4
White	3
Do not like any chocolate	2

   (a) How many pupils are in his class?

   (b) What percentage of the class is one pupil?

   (c) What percentage of the class preferred plain chocolate?

   (d) Work out all the percentages and draw the pie chart. Label each sector.

3 In a survey 200 young people were asked about which sport they liked to watch on television.

The information is shown in the table.

Football	Motor racing	Snooker	Tennis	Cricket
90	64	12	24	10

   (a) Draw a pie chart to illustrate this information. Label each sector clearly.

Some senior citizens were also asked the same question.
This pie chart shows the results.

   (b) Give one way in which the results for young people and for senior citizens are similar.

   (c) Give one way in which the senior citizens' results are different from the young people's results.

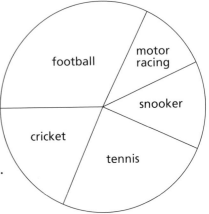

4 A company runs four types of Family Activity Holiday.

The table shows the number of each type of holiday taken in one month.

Draw and label a pie chart to show this information. Show your working.

Type of holiday	Frequency
Skiing	115
Cycling	50
Walking	40
Boating	45

Section F

1 The table shows how 600 people left the United Kingdom to travel abroad.

Draw and label a pie chart to illustrate this data.

Method of transport	Frequency
Channel tunnel	84
Sea	151
Air	365
Total	**600**

2 Gurpreet asked his class which type of paper they liked to use for maths notes and activities.

Squared	Wide lined	Narrow lined	Plain
8	13	4	3

Draw a pie chart to show this data. Label your chart.

3 The table shows the amount of certain materials that were recycled in 1998–1999 in the United Kingdom.

Draw and label a pie chart to show this information.

Material	Tonnes (thousands)
Paper and card	835
Compost	525
Glass	363
Co-mingled material	119
Cans	32
Other	390
Total	**2264**

Mixed questions 6

1 This is a scale drawing of a bedroom. The scale of the drawing is 2 cm to 1 metre.

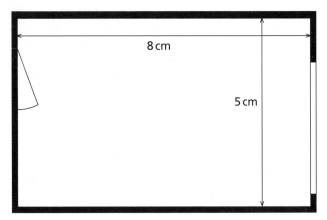

   (a) What is the length of the real bedroom?

   (b) What is the width of the real bedroom?

   (c) The bed is 2 m long. How long should it be on the scale drawing?

2 How much is left when you use

   (a) 250 grams of sugar from a 1 kilogram bag

   (b) 100 millilitres of vinegar from a $\frac{1}{2}$ litre bottle

   (c) 15 centimetres of ribbon from a 2 metre roll

3 Roughly how many litres are there in

   (a) 2 pints   (b) 2 gallons   (c) 12 pints

4 This line graph shows the temperatures in °C at noon in Bombay over one week.

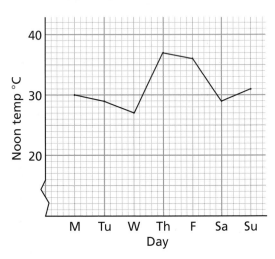

   (a) What was the noon temperature on Monday?

   (b) What was the noon temperature on Saturday?

   (c) Between which two days did the temperature go up most?

5 This table shows the midnight temperatures in Bombay over the same week.

Day	M	Tu	W	Th	F	Sa	Su
Midnight temp (°C)	18	19	25	21	20	26	22

On graph paper draw a line graph to show this data.

6 Work out the value of each expression when $a = 2$, $b = 3$ and $c = 4$.

(a) $5b$ (b) bc (c) $\dfrac{bc}{3}$ (d) $1 + \dfrac{c}{a}$ (e) $5 - \dfrac{a}{c}$

7

*I think of a number.
I add 3 to the number.
I multiply my result by 5.*

Look at this number puzzle.

(a) Sofima starts with 4.
   What is her answer?

(b) Write down an expression for the answer if the number you start with is n.

8

A	B	C	D	E	F	G	H	I	J	K	L
⁻5	⁻4	⁻3	⁻2	⁻1	0	1	2	3	4	5	6

Work out the value of each expression below when $p = {}^-2$, $q = 4$ and $r = {}^-8$.
Then use the code above to turn each value into a letter.
In each part rearrange the letters to make a girl's name.

(a) $r + 11$ $q - 7$ $p + 6$ $p + 7$ $r + 3$ $q - 5$

(b) $1 - p$ $p - r$ $2q - 2$ $2p + 3$ $r + 7$

(c) $\dfrac{r}{4}$ $\dfrac{p}{2}$ $4 - 2q$ $\dfrac{r}{q} - 2$ $\dfrac{r}{p} - 5$ $1 - \dfrac{q}{p}$

(d) $pr - 10$ $3 + pq$ $p^2 - 9$ $q^2 - 13$ $12 - q^2$ $\dfrac{p^2}{4}$ $\dfrac{r^2}{16} - 1$

9 Work out as percentages

(a) 68 out of 200   (b) 16 out of 200   (c) 56 out of 175   (d) 9 out of 225

10 Work out each of these. Give your answers to the nearest penny.

(a) 17% of £24.50   (b) 38% of £49.70   (c) 7% of £12.65   (d) 15% of £37.77

11 The diagram shows five cubes joined together.

Write down the coordinates of A, B and C.

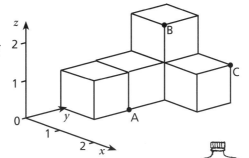

12 *Value* orange squash comes in three sizes.

Which size bottle works out cheapest?
Explain your answer.

500 ml   1.2 litre   3 litre
72p   £1.50   £4.20

13 (a) Hema bought 3 paint brushes and 4 tins of blue paint.
The brushes cost £1.99 each, and the total bill was £23.05.
How much did each tin of paint cost?

(b) The next day she bought 2 tins of varnish and some tins of paint remover.
The varnish cost £1.89 a tin, and the paint remover cost £1.59 a tin.
She got 9p change from £15.00.
How many tins of paint remover did she buy?

14 Copy and complete

(a) $2(x + 3) = 2x + \blacksquare$ (b) $3(n - 4) = \blacklozenge - 12$ (c) $\blacktriangledown(h + 3) = 4h + 12$

15 Factorise each of these as far as you can.

(a) $12n + 24$ (b) $6g - 12$ (c) $18s + 24$ (d) $12d - 16$

16 Simplify these.

(a) $2(x + 3) + 5$ (b) $3(d + 1) + 2d$ (c) $5(w - 2) + 12$ (d) $8 + 2(w - 1)$

17 This pie chart gives information about
the audience for a science fiction film.

(a) What fraction of the audience was men?

(b) What fraction was women?

(c) Calculate the size of the angle for boys.
What fraction of the audience was boys?

(d) There were 300 people in the
audience altogether.
How many girls were in the audience?

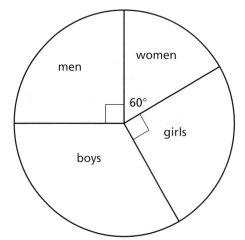

18 Rhys asked 40 people what they
drank with their breakfast.
This pie chart shows his results.

(a) Use the pie chart to work out how many
people drank tea with their breakfast.

(b) Work out how many people
did not drink coffee.

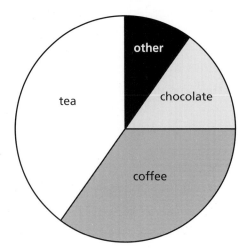

53 Patchwork

Section A

1 What fraction of each design is black?
 Give each answer in its simplest form.

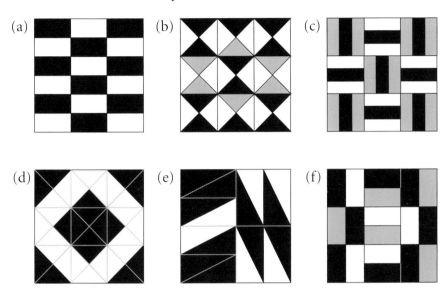

(a)   (b)   (c)
(d)   (e)   (f)

2 Design your own pattern where $\frac{1}{4}$ of the design is black.

Section B

1 Mike is laying a patio.
 Here is one block of his design.

 (a) What fraction of the block is black?

 (b) What fraction of the block is grey?

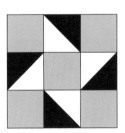

2 Mike needs 25 of these blocks.

 (a) How many grey slabs will be needed?

 (b) What fraction of the patio will be white?

3 Here is a block of a floor tile design.

 (a) What fraction of this block is grey?

 (b) What fraction of this block is white?

 (c) What fraction of this block is black?

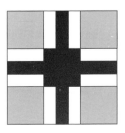

4 (a) What fraction of this design is

    (i) black

    (ii) white

  (b) If the design is a 12 cm by 12 cm square, what is the area of the black part of the design in square centimetres?

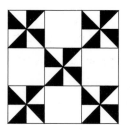

5

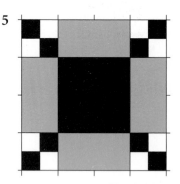

  (a) What fraction of this design is

    (i) grey

    (ii) white

    (iii) black

  (b) If a small white square is 1 cm by 1 cm, find the area of the black part of the design in square centimetres.

Section C

1 Copy the following designs, mark in any lines of symmetry with a dotted line. Mark any centre of rotation and write the order of rotation symmetry under each one.

(a)

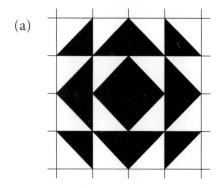

(b)

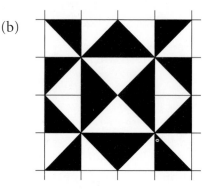

(c)

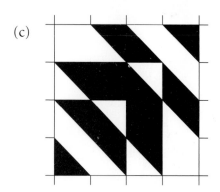

(d)

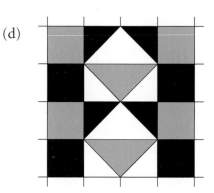

172

2 In the diagrams below, lines of symmetry are shown by dotted lines.
Centres of rotation are shown by black dots.

Copy and complete each block.

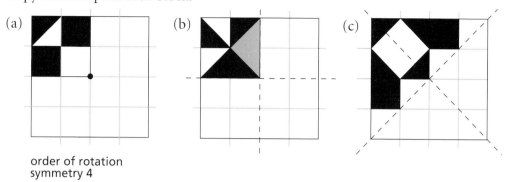

(a)

order of rotation
symmetry 4

(b)

(c)

Section E

1 Work out the size of each angle marked with a letter.

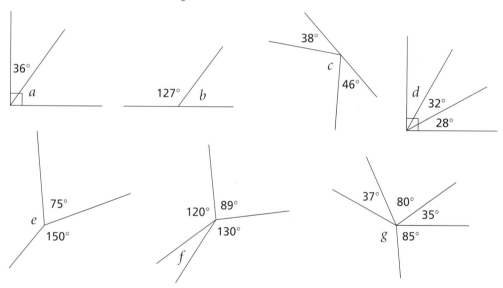

2 Work out the size of each angle marked with a letter.

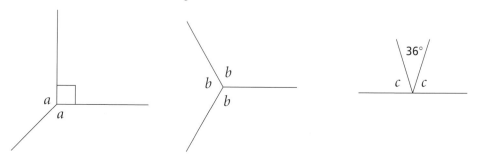

Section F

1 Work out the size of each angle marked with a letter.

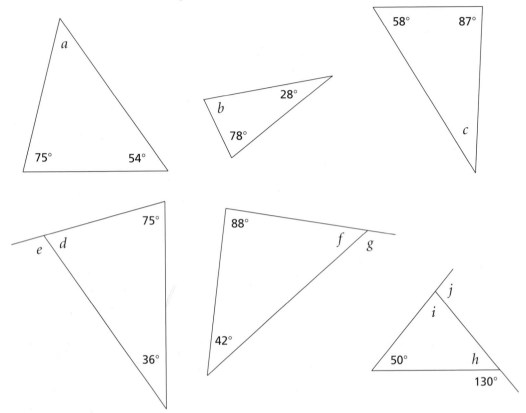

2 Work out the size of each angle marked with a letter.

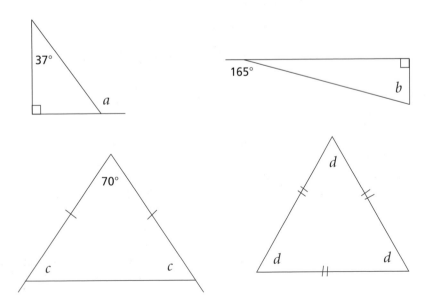

54 *Travel*

Sections A, B and C

1 Change these times into 24-hour clock times.

   (a) 4.15 p.m.   (b) half past six in the morning

   (c) 12.40 a.m.   (d) quarter to seven in the evening

2 This is Jayne's school timetable for a Monday.

   (a) Write the time school finishes
using a.m. or p.m.

   (b) How long is morning registration?

   (c) How long are morning lessons?

   (d) How long is her afternoon lesson?

   (e) How long does Jayne's school day last?

   (f) Jayne's mum says she will pick her up at a
quarter to four after school.
How long will Jayne have to wait?

08:35	Registration
08:50	Assembly
09:10	Maths
10:05	PE
11:00	Morning break
11:20	French
12:15	Lunchtime
13:25	Registration
13:30	Design and Technology
15:20	School finishes

3 Here is part of the timetable for trains
on the North Yorkshire Moors Railway.

   (a) What time does the 12:20 train
from Pickering arrive in Grosmont?

   (b) How long do trains take to get
from Pickering to Newton Dale?

Pickering	11:20	12:20	14:20	15:20
Levisham	11:40	12:40	14:40	15:40
Newton Dale	11:49	12:49	14:49	15:49
Goathland	12:10	13:10	15:10	16:10
Grosmont	12:25	13:25	15:25	16:25

   (c) Meena wants to arrive in Goathland by 2 p.m.
What is the latest train she can get from Pickering?

4 This table shows the distances, in miles, between some English cities.

   (a) How far is it between Exeter and London?

   (b) How far is it between Cambridge and Oxford?

   (c) Dennis drives from Cambridge to Exeter.
He then drives from Exeter to Bristol and
back to Cambridge.
How far has he driven altogether?

Bristol	Cambridge	Exeter	London	Oxford	Southampton
171					
83	250				
120	60	200			
74	100	154	56		
76	131	112	80	66	

Section D

1 Find the average speed, in metres per second, of

   (a) a sailfish swimming 900 m in 30 seconds

   (b) a cheetah running 1450 m in 50 seconds

   (c) a barnacle goose flying 360 m in 20 seconds

2 Work out the average speed of these in km/h.

   (a) A train that goes 240 km in 3 hours

   (b) A plane that flies 2120 km in 4 hours

   (c) A spacecraft that travels 105 000 km in 5 hours

3 Calculate the average speed for each of these journeys.
   State the units clearly in your answer.

   (a) A motorbike that goes 350 miles in 5 hours

   (b) A lion running 264 metres in 12 seconds

   (c) A boat that sails 6 km in 20 minutes

Section E

1 Azmat went for a cycle ride. This graph shows his journey.

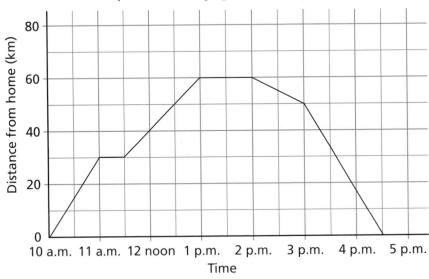

   (a) What was Azmat's average speed for the first hour?

   (b) What happened at 11 a.m.?

   (c) Between which two times was Azmat cycling slowest?

   (d) When did he start his journey home?

   (e) How far did he cycle altogether?

2 Draw a set of axes with the same scales
as in question 1 but with the time going to 6 p.m.

   (a) Draw the graph to show this journey.

      • We left home at 10 a.m. and cycled at an average speed of 25 km/h for two hours.

      • In the next hour we went faster and covered 30 km.

      • We then stopped for lunch for an hour.

      • After lunch we headed back and arrived back home at 6 p.m.

   (b) What was the average speed on the return journey?

Section F

1 A McLaren F1 can drive at a maximum speed of 240 m.p.h.

   How far could it travel in

   (a) half an hour   (b) 45 minutes

2 Connor can walk at a steady speed of 5 km/h.

   How long does he take to walk 20 km?

3 How far are each of these journeys?

   (a) A hawkmoth flies at 14 m/s for 30 seconds.

   (b) A greyhound runs at 17 m/s for 20 seconds.

   (c) A plane travels at 550 km/h for 3 hours.

   (d) A man runs at 8 km/h for 15 minutes.

   (e) A bus travels at 25 m.p.h. for 30 minutes.

4 How long would each of these journeys take?

   (a) A man walking at 6 km/h for 3 km

   (b) A woman cycling at 25 km/h for 100 km

   (c) A boat travelling at 6 km/h for 15 km

   (d) A motorbike travelling at 52 m.p.h. for 39 miles

Section G

1 Work out the average speeds of these journeys.

   (a) Rita drives from Birmingham to Swindon in 2 hours 15 minutes.
        This is a distance of 90 miles.

   (b) Martin flies from Gatwick to Glasgow, a distance of 445 miles.
        The journey takes $2\frac{1}{2}$ hours.

   (c) Stacey gets a coach from Aberdeen to Hull, a distance of 360 miles.
        The journey takes 7 hours 30 minutes.

2 Maxine went for a walk.
   Her walk is represented by the graph below.

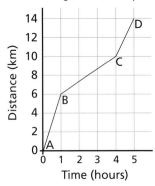

   (a) What was her average speed on section AB?

   (b) What was her average speed on section CD?

   (c) On which section of the walk did she walk the slowest?

3 In January 2000, Jeremy Wetherspoon skated 500 metres in just 34.63 seconds.
   What was his average speed in m/s correct to two decimal places?

4 Cathy left home in her car at 08:30.
   She arrived at the motorway at 09:15.

   (a) It is 15 miles from her home to the motorway.

        What was her average speed for this part of the journey?

   (b) Cathy then drove along the motorway at a steady speed of 68 miles per hour.
        She had to drive for 85 miles along the motorway.

        For how long was she on the motorway?

   (c) Cathy drove home by a different route.
        Her average speed was 54 miles per hour.
        The journey home took her $2\frac{1}{2}$ hours.

        What distance was the journey home?

55 Cuboids

Section A

1 For each of these boxes work out how many 1 cm cubes it would take to

    (i) make a layer to cover the bottom

    (ii) completely fill it

(a)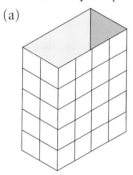

(b)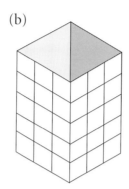

(c)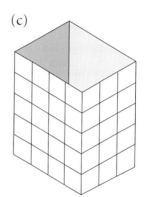

2 Find the volume of these cuboids.

(a)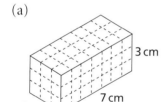

(b)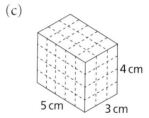

(c)

(d)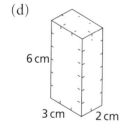

(e)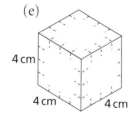

(f)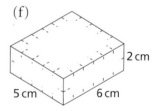

(g)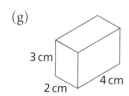

(h)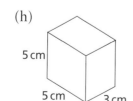

(i)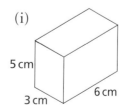

3 A printer cartridge carton measures 10 cm by 5 cm by 5 cm.

   (a) What is the volume of a printer cartridge carton?

   40 printer cartridge cartons are packed in a box.

   (b) What is the volume of the box?

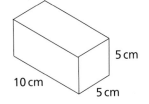

4 Find the volume of these cuboids, stating your units.

   (a) height = 8 cm, length = 9 cm, width = 5 cm

   (b) height = 10 cm, length = 6 cm, width = 4 cm

5 Find the volume of each of these packets.

(a) 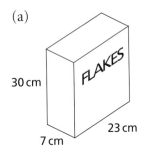 (b) 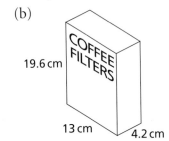 (c)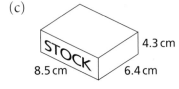

6 Find the volume of these cuboids, stating your units.

   (a) height = 4.5 cm, length = 12.6 cm, width = 4.8 cm

   (b) height = 0.8 cm, length = 3.8 cm, width = 0.9 cm

7 A cuboid-shaped packet containing 10 textbooks measures 25 cm by 20 cm by 20 cm.

   (a) What is the volume of the packet?

   (b) What is the volume of each textbook?

8 These cuboids all have a volume of 60 cm^3.

   Find the measurements labelled with letters.

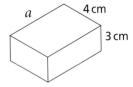 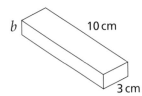 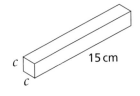

9 400 cubes each with a volume of 1 cm^3 will fill each of these boxes.

   What is the height of each box?

   (a) length = 10 cm, width = 8 cm, height = ?

   (b) length = 16 cm, width = 5 cm, height = ?

   (c) length = 20 cm, width = 5 cm, height = ?

Sections B and C

1 Find the volume of these cuboids in m³.

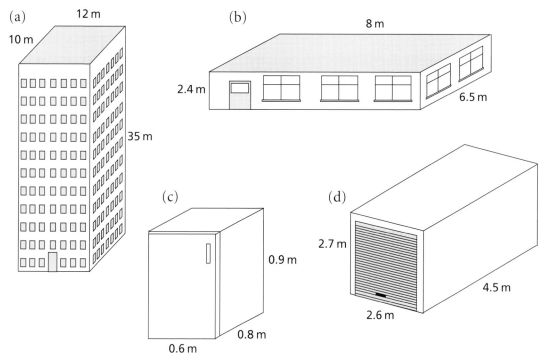

(a) 12 m, 10 m, 35 m

(b) 8 m, 2.4 m, 6.5 m

(c) 0.9 m, 0.8 m, 0.6 m

(d) 2.7 m, 4.5 m, 2.6 m

2 A store room measures 3 m long, 4 m wide and 2.5 m high.

  (a) Calculate the volume of the store room in m³.

  (b) Write down the measurements of the store room in centimetres.

  (c) Use these to find the volume of the store room in cm³.

3 The measurements of this paving stone are 60 cm by 60 cm by 4 cm.

  They can be bought in packs of 20 stones.

  4 cm   60 cm   60 cm

  (a) Write down the height of the pack of stones in centimetres.

  (b) Find the volume of the pack of stones in cm³.

  (c) Write down the measurements of the pack of stones in metres.

  (d) Find the volume of the pack in m³.

Section D

1 This is the net of a cuboid.

(a) What are the measurements
of the cuboid?

(b) Calculate the volume of
the cuboid.

(c) Calculate the surface area of
the cuboid.

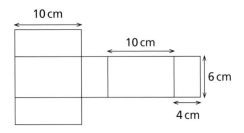

2 Draw sketches of the nets of these cuboids.

Use the nets to calculate the surface area of the cuboids.

(a)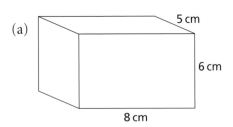

(b)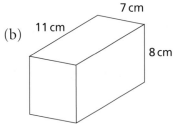

3 The diagrams show some blocks of wood.

Find the surface area of each block of wood.

It may help to draw nets of the solids.

(a)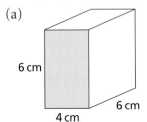

(b)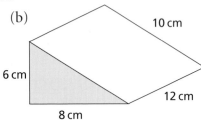

(c)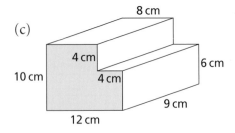

(d)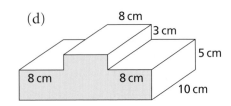

56 *More circle facts*

Sections A and B

1 Draw a circle and label clearly its radius, diameter and circumference.

2 Find the rough circumference of these circles.

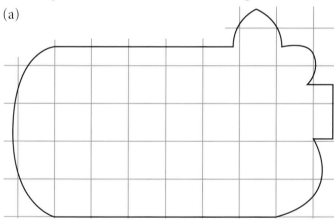

(a) 4 cm

(b) 10 cm

(c) 6 cm

(d) ←—8 cm—→

Not to scale

3 A hamster wheel has a diameter of 15 cm.

(a) Use π to find the circumference of the wheel.

(b) Hammy Hamster does 10 rotations of the wheel. How far has he run?

4 These patches for jeans are drawn on centimetre squared grids.
Use the grid to estimate the area of each patch.

(a)

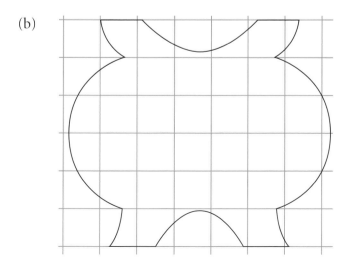

(b)

183

Section C

1 Write down the formula for finding the area of a circle using π.

2 Find the areas of these circles.

(a)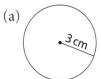
(b)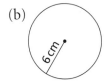
(c)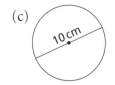
(d)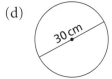

Not to scale

3 Here are some pictures of foil bottle tops.

(a)   (b)   (c)

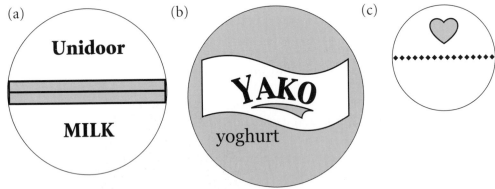

(i)   Measure the diameter of each top to the nearest 0.1 cm.

(ii) Work out the radius of each top.

(iii) Calculate the area of each top.

4 Annie is trying out different logo designs for a company.
   Her designs are based on these basic shapes.

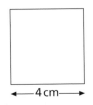

(a) Find the area of each of the basic shapes.

(b) Find the area of each of these designs, which use the shapes above.

(i) 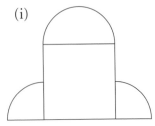   (ii)    (iii)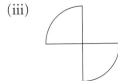

Section D

1 Table tennis balls are spherical.
Each ball has a diameter of 4 cm.
Four balls fit tightly in a box as shown.

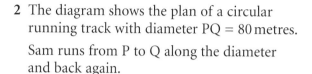

(a) What is the length, width and height of the box?

(b) What is the volume of the box? State your units.

2 The diagram shows the plan of a circular
running track with diameter PQ = 80 metres.

Sam runs from P to Q along the diameter
and back again.

Tom runs from P around the circumference of
the track and back to P again.

(a) How far does each person run?

(b) Who runs further and by how much?

3 This design is for a sports logo.

(a) Measure accurately the diameters
of the circles.

(b) Calculate the shaded area.
Show all your workings.

4 Susan makes this pendant in white and black enamel.

(a) Calculate the area of black enamel used.

(b) Calculate the area of white enamel used.

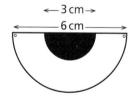

5 John is making a sculpture from metal rods.
He needs to make this piece, which is a quarter of a circle.
What length of rod does he need?

185

57 Interpreting data

Sections A and B

1 Tony carried out a survey to see how many cars passed his house.
   Here are the results of the first five one-minute intervals.

   Number of cars: 6, 3, 7, 6, 3

   Find

   (a) the range   (b) the median   (c) the mean

2 In the 2001–2002 football season, the numbers of goals scored
   by seven footballers at Munchester Town were

      30, 17, 10, 19, 16, 13, 14

   In the 2002–2003 season, the numbers of goals scored by the same seven footballers
   were

      14, 15, 11, 15, 19, 10, 14

   (a) Calculate the mean number of goals scored in the 2001–2002 season.

   (b) Calculate the mean number of goals scored in the 2002–2003 season.

   (c) Find the range of the goals scored in the 2001–2002 season.

   (d) Find the range of the goals scored in the 2002–2003 season.

   (e) Make two comments about the goal scoring of the seven footballers
       in the two seasons.

3 In a memory test some girls were shown 20 objects for 20 seconds.
   They were then given one minute to list as many of the objects as possible.

   The results of the test are shown in the table.

   (a) How many girls took the test?

   (b) What is the range of the number of
       items remembered?

   (c) What is the modal number of
       items remembered?

   (d) Find the median number of items remembered.

   (e) Draw a frequency diagram to show the
       information in the table.

Items remembered	Frequency
5	1
6	2
7	4
8	10
9	8
10	6
11	3
12	1

Section C

1 The data shows the marks for the first paper of a mathematics examination.

56	74	51	33	50	60	49	42	52	66	39	45	41	51	66
71	69	44	56	59	84	66	53	37	65	40	53	45	50	

(a) Put this data into a stem-and-leaf table, using this stem.

(b) Find the median mark and the range.

The data below shows the marks for the second paper.

62	73	47	54	53	68	61	75	74	61	28	72	48	56	74
81	59	78	74	57	71	72	64	62	71	67	42	55	38	

(c) Put this data into a stem-and-leaf table using the same stem.

(d) Find the median and the range.

(e) Write a comment on the difference between the two papers.

2 (a) Make a frequency table for the data for the first mathematics paper in question 1 above.

Use groups such as 20–29 and 30–39.

(b) What is the modal group for the first paper?

(c) Draw a frequency diagram showing the marks.

(d) Using the same groups, what is the modal group for the second paper?

3 The data below shows some children's estimates of the number of fish in a pond.

10	8	20	10	8	9	8	10	15	9	10	10	12	15
15	10	8	5	12	9	10	14	8	9	10	15	10	16

(a) Using groups 5–7, 8–10, 11–13, ... make a frequency table for the estimates.

(b) Draw a frequency diagram using your table.

Section D

1 To see how good they were at estimating, 30 children tried to draw a line 25 cm long without a ruler.

The data below shows the actual lengths of the lines.

24.8 22.0 30.1 25.0 19.6 27.3 25.9 28.1 21.2 25.7 32.0 22.9 20.6 30.8 30.1
30.6 22.5 19.1 27.6 25.4 21.8 27.8 26.2 20.0 25.8 23.0 24.3 27.6 20.4 31.1

(a) Copy and complete this grouped frequency table to record the lengths.

(b) What is the modal group of lengths?

(c) How many children drew lines less than 25 cm long?

(d) Draw a frequency graph for the lengths using these scales.

Length l (cm)	Frequency
$19.0 \leq l < 22.0$	
$22.0 \leq l < 25.0$	
$25.0 \leq l < 28.0$	
$28.0 \leq l < 31.0$	
$31.0 \leq l < 34.0$	
Total	

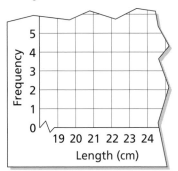

(e) Draw a frequency polygon on your graph.

2 The distances, in metres, jumped by seventeen triple jumpers in a competition are shown below.

10.9   11.4   14.2   13.7   11.0   12.8   12.7   11.8   12.1   13.3
13.8   13.1   12.6   12.2   12.5   13.1   12.4

(a) Copy and complete this frequency table for the 17 jumps.

(b) What is the modal group?

(c) Draw a frequency graph for the jumps.

(d) Draw a frequency polygon on your graph.

Jump j (m)	Frequency
$10.0 \leq j < 11.0$	
$11.0 \leq j < 12.0$	
$12.0 \leq j < 13.0$	
$13.0 \leq j < 14.0$	
$14.0 \leq j < 15.0$	
Total	

58 Sequences

Section A

1 For each of these sequences the first four terms and the rule is given.
 Write down the next three terms of each sequence.
 (a) 1   5   9   13 ...   Add 4 to the last term.
 (b) 1   2   4   8 ...   Double the last term.
 (c) 1   4   13   40 ...   Multiply the last term by 3 and add 1.

2 Work out the next two terms in each of these sequences.
 Write down the rule for finding the next term in your sequence.
 (a) 3   8   13   18 ...
 (b) 6   12   24   48 ...
 (c) 1   3   4   7   11 ...

3 Work out the next two terms in each sequence.
 Write down the rule for finding the next term.
 (a) 11   8   5   2   ⁻1 ...
 (b) 12   7   2   ⁻3 ...
 (c) 5 $4\frac{1}{2}$ 4 $3\frac{1}{2}$ ...

4 David has this number pattern.
      5   10   20   40 ...
 (a) Explain how to work out the next number in David's pattern.
 (b) Write down the next two numbers in this pattern.

5 In this number pattern, you add the last two numbers to get the next number.
      4   5   9   14 ...
 Write down the next two numbers in this pattern.

*6 A sequence starts 1   3   7 ...
 Jo says the next two terms are 13   21.
 Peter says the next two terms are 15   31.
 Both of them are right!
 (a) Explain what Jo's rule is.   (b) Explain what Peter's rule is.

Section B

1 Here is a pattern made with matches.

  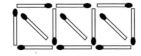 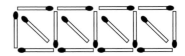

| Pattern 1 | Pattern 2 | Pattern 3 |

(a) Copy and complete the table.

Pattern	1	2	3	4	5
Number of matches	9	13			

(b) Which pattern can be made with exactly 33 matches?

(c) Explain how you could work out the number of matches needed for pattern 10 without doing any drawing.

2 Here are some patterns of dots.

| Pattern 1 | Pattern 2 | Pattern 3 |

(a) Draw pattern 4.

(b) Copy and complete the table.

Pattern	1	2	3	4	5
Number of dots	6	10			

(c) Write down the number of dots needed for pattern 10.
Explain how you found your answer.

3 Shape 1　　Shape 2　　　　Shape 3

 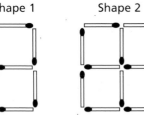 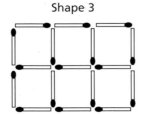

(a) Copy and complete this table for the shapes above.

Shape	1	2	3	4	5
Number of matches	7				

(b) How many matches would there be in shape 15?
Explain how you got your answer.

(c) One shape is made from 102 matches.
Which shape is this?

190

59 Enlargement

Sections B and C

1 (a) Carefully copy these shapes on to centimetre squared paper.

(b) Draw extended lines between matching points on the shapes to find the centre of enlargement.

(c) Mark the centre of enlargement with an X.

(d) Write down the scale factor of the enlargement.

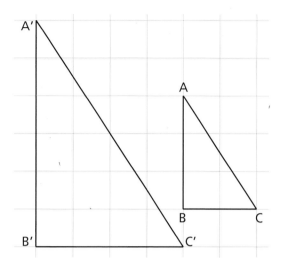

2 Follow the same steps as in question 1 with these two shapes.

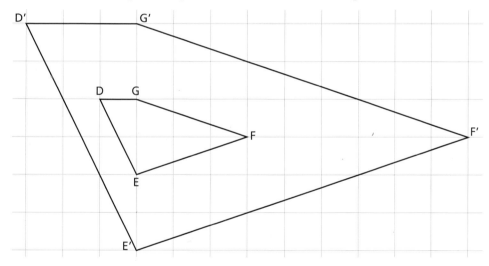

3 Copy this shape on to centimetre squared paper.

Draw an enlargement of the shape with scale factor 2, centre O.

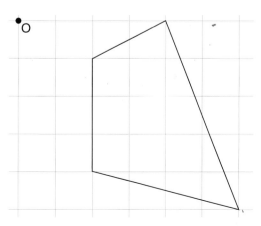

4 Copy this hexagon and the point X on to centimetre squared paper.

Draw an enlargement of the shape with scale factor 2 and centre X.

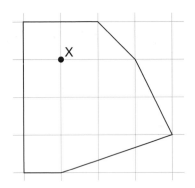

5 Find the scale factor of the enlargement from

(a) shape A to shape B

(b) shape A to shape C

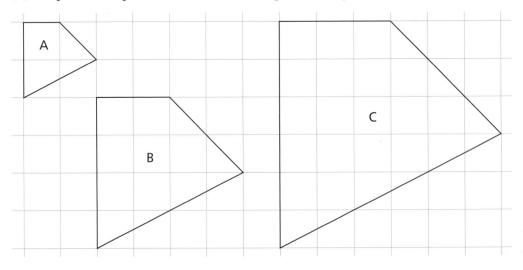

6 A rectangle has width 4 cm and length 6 cm.
It is enlarged by scale factor 4.
What will be the width and length of the enlarged rectangle?

7 A quadrilateral has angles of 48°, 99°, 66° and 147°.
The perimeter of the quadrilateral is 45 cm.
The quadrilateral is enlarged by scale factor 2.

(a) What are the angles of the enlarged quadrilateral?

(b) What is the perimeter of the enlarged quadrilateral?

60 *Calculating with fractions*

Section B

1 Match each mixed number with an improper fraction.

$\frac{8}{3}$ $\frac{9}{5}$ $\frac{7}{3}$ $\frac{6}{5}$ $\frac{16}{5}$ $1\frac{1}{5}$ $2\frac{2}{3}$ $3\frac{1}{5}$ $1\frac{4}{5}$ $2\frac{1}{3}$

2 Change these mixed numbers into improper fractions.
   (a) $\frac{13}{5}$ (b) $\frac{5}{3}$ (c) $\frac{12}{7}$ (d) $\frac{13}{6}$ (e) $\frac{19}{8}$

3 Arrange these fractions in order of size, smallest first.

   $\frac{14}{3}$ $\frac{13}{6}$ $\frac{12}{7}$ $\frac{13}{4}$ $\frac{10}{5}$

Section C

1 Janet only likes red or black chewy sweets.
   $\frac{2}{7}$ of a box of chewy sweets are red and $\frac{1}{7}$ of the box are black.

   What fraction of the box of sweets does Janet like?

2 Work these out, simplifying the answer where possible.
   (a) $\frac{2}{5} + \frac{1}{5}$ (b) $\frac{5}{6} - \frac{1}{6}$ (c) $\frac{3}{4} - \frac{1}{4}$ (d) $\frac{3}{7} + \frac{3}{7}$ (e) $\frac{5}{8} + \frac{1}{8}$
   (f) $\frac{4}{9} + \frac{2}{9}$ (g) $\frac{5}{9} - \frac{2}{9}$ (h) $\frac{3}{10} + \frac{1}{10}$ (i) $\frac{7}{10} - \frac{2}{10}$ (j) $\frac{5}{12} + \frac{3}{12}$

3 Simon spent $\frac{3}{10}$ of his holiday money on the first day of his holiday and $\frac{1}{10}$ of his holiday money on the second day.
   (a) What fraction of his holiday money did Simon spend on the first two days?
   (b) What fraction did he have left for the remainder of his holiday?

4 Work these out, simplifying your answers where possible.
   (a) $\frac{3}{5} + \frac{2}{5}$ (b) $\frac{8}{9} + \frac{4}{9}$ (c) $1\frac{1}{4} + \frac{3}{4}$ (d) $1 - \frac{3}{5}$ (e) $1\frac{3}{8} - \frac{5}{8}$
   (f) $\frac{5}{8} + 2\frac{3}{8}$ (g) $2\frac{4}{7} - \frac{3}{7}$ (h) $2\frac{4}{7} + 1\frac{5}{7}$ (i) $1\frac{3}{10} - \frac{7}{10}$ (j) $2\frac{1}{10} - 1\frac{9}{10}$

5 Work these out, simplifying your answers where possible.
   (a) $\frac{1}{2} \times 10$ (b) $\frac{1}{4} \times 7$ (c) $\frac{2}{3} \times 4$ (d) $\frac{3}{4} \times 5$ (e) $\frac{4}{5} \times 6$
   (f) $1\frac{1}{2} \times 3$ (g) $2\frac{1}{4} \times 4$ (h) $3\frac{1}{3} \times 5$ (i) $1\frac{2}{9} \times 6$ (j) $2\frac{3}{10} \times 5$

6 Amit works for $2\frac{3}{4}$ hours a day, five days a week.
   How many hours does he work in a week?

193

Sections D and E

1 For each pair of fractions, write down the larger fraction.

(a) $\frac{3}{5}, \frac{2}{5}$ (b) $\frac{1}{3}, \frac{1}{4}$ (c) $\frac{2}{5}, \frac{2}{7}$ (d) $\frac{1}{3}, \frac{4}{9}$ (e) $\frac{5}{8}, \frac{9}{16}$

2 Put each set of fractions in order starting with the smallest.
What word do you make for each set of fractions?

(a)

$\frac{1}{2}$	$\frac{3}{4}$	$\frac{2}{3}$	$\frac{5}{6}$	$\frac{7}{12}$
D	A	E	M	R

(b)

$\frac{4}{5}$	$\frac{7}{10}$	$\frac{3}{5}$	$\frac{9}{10}$	$\frac{1}{2}$
C	A	R	E	B

(c)

$\frac{3}{4}$	$\frac{3}{8}$	$\frac{1}{2}$	$\frac{5}{8}$	$\frac{9}{16}$
R	T	I	E	M

3 Work these out, simplifying your answers where possible.

(a) $\frac{1}{2} + \frac{1}{8}$ (b) $\frac{1}{6} + \frac{1}{12}$ (c) $\frac{3}{4} - \frac{1}{2}$ (d) $\frac{3}{10} + \frac{1}{5}$ (e) $\frac{7}{10} - \frac{2}{5}$

(f) $\frac{3}{8} - \frac{1}{4}$ (g) $\frac{2}{3} + \frac{1}{12}$ (h) $\frac{7}{12} - \frac{1}{3}$ (i) $\frac{4}{15} + \frac{1}{3}$ (j) $\frac{7}{9} - \frac{2}{3}$

4 Martin cut $\frac{1}{3}$ of his hedge in the morning and $\frac{1}{6}$ in the afternoon.
How much of his hedge did he cut altogether?

5 A cylinder is $\frac{1}{4}$ full of water.
Another cylinder, the same shape and size, is $\frac{3}{8}$ full.
The water from the first cylinder is poured into the
second cylinder.

How full is the second cylinder now?

$\frac{1}{4}$ $\frac{3}{8}$

6 Work these out, giving your answers as mixed numbers, simplified where possible.

(a) $\frac{3}{4} + \frac{5}{8}$ (b) $1\frac{7}{10} - \frac{3}{5}$ (c) $2\frac{1}{4} - \frac{1}{2}$ (d) $1\frac{1}{4} + \frac{5}{8}$ (e) $\frac{11}{12} + \frac{1}{3}$

(f) $\frac{5}{9} + \frac{2}{3}$ (g) $2\frac{1}{3} - \frac{5}{6}$ (h) $\frac{5}{12} + \frac{3}{4}$ (i) $3\frac{1}{2} - 1\frac{3}{4}$ (j) $1\frac{3}{4} + \frac{3}{8}$

7 Monty drank $\frac{3}{5}$ litre of water before lunch and $\frac{7}{10}$ litre after lunch.

(a) How much water did Monty drink?

(b) He wanted to drink 2 litres. How much more does he need to drink?

Sections F and G

1 Below are sets of equivalent fractions for $\frac{3}{4}$ and $\frac{5}{6}$.

$$\frac{3}{4} = \frac{6}{8} = \frac{9}{12} = \frac{12}{16} = \frac{15}{20}$$

$$\frac{5}{6} = \frac{10}{12} = \frac{15}{18} = \frac{20}{24} = \frac{25}{30}$$

Use the lists to decide which is larger, $\frac{3}{4}$ or $\frac{5}{6}$.
Explain how you decided.

2 Arrange these sets of fractions in ascending order. What words do you make?

(a)

$\frac{3}{5}$	$\frac{4}{5}$	$\frac{3}{4}$	$\frac{1}{2}$
O	T	S	H

(b)

$\frac{5}{12}$	$\frac{1}{2}$	$\frac{1}{4}$	$\frac{1}{3}$
T	S	A	R

(c)

$\frac{3}{5}$	$\frac{2}{3}$	$\frac{8}{15}$	$\frac{1}{2}$
L	A	O	C

3 Use these sets of equivalent fractions to work out $\frac{3}{4} + \frac{2}{5}$.

$$\frac{3}{4} = \frac{6}{8} = \frac{9}{12} = \frac{12}{16} = \frac{15}{20} = \frac{18}{24} = \frac{21}{28}$$

$$\frac{2}{5} = \frac{4}{10} = \frac{6}{15} = \frac{8}{20} = \frac{10}{25} = \frac{12}{30} = \frac{14}{35}$$

4 (a) Copy and complete these sets of equivalent fractions.

(i) $\frac{1}{4} = \frac{2}{8} = \frac{\blacksquare}{12} = \frac{4}{\blacksquare} = \frac{\blacksquare}{\blacksquare}$

(ii) $\frac{2}{3} = \frac{4}{\blacksquare} = \frac{\blacksquare}{9} = \frac{\blacksquare}{\blacksquare} = \frac{\blacksquare}{\blacksquare}$

(b) Use your answers to work out $\frac{1}{4} + \frac{2}{3}$.

5 Work these out, simplifying your answers where possible.

(a) $\frac{1}{2} + \frac{1}{3}$ (b) $\frac{1}{6} + \frac{1}{4}$ (c) $\frac{3}{5} - \frac{1}{3}$ (d) $\frac{1}{5} + \frac{1}{4}$ (e) $\frac{2}{5} - \frac{1}{3}$

(f) $\frac{3}{7} - \frac{1}{4}$ (g) $\frac{2}{5} + \frac{1}{2}$ (h) $\frac{5}{6} - \frac{1}{4}$ (i) $\frac{7}{15} + \frac{1}{10}$ (j) $\frac{7}{8} - \frac{5}{6}$

6 $\frac{1}{5}$ of the children in a school travel to school by bus.

$\frac{1}{3}$ of the children travel to school by car.

What fraction travel to school by bus or by car?

7 A sheet of plastic $\frac{5}{12}$ inch thick is stuck
to a plywood sheet $\frac{7}{8}$ inch thick.
What is the total thickness of the two sheets?

Plastic ($\frac{5}{12}$ inch)

Plywood ($\frac{7}{8}$ inch)

61 Substitution

Sections A and B

1 Work out each of these.

    (a) $5 \times 3 + 2$ (b) $5 \times (3 + 2)$ (c) $5 + 3 \times 2$

2 Find **three** pairs that give the same answer.

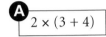 **A** $2 \times (3 + 4)$ 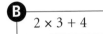 **B** $2 \times 3 + 4$ 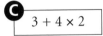 **C** $3 + 4 \times 2$ 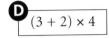 **D** $(3 + 2) \times 4$

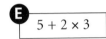 **E** $5 + 2 \times 3$ 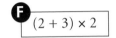 **F** $(2 + 3) \times 2$  **G** $2 + 3 \times 2$  **H** $6 + 1 \times 2$

3 Copy these and put in brackets to make them correct.

    (a) $2 + 3 \times 4 + 5 = 25$ (b) $2 + 3 \times 4 + 5 = 45$

4 Work out these expressions when $f = 3$, $g = 6$ and $h = 9$.

    (a) fg (b) $fg + h$ (c) $f(g + h)$ (d) $\dfrac{h + g}{f}$

    (e) $f^2 + g^2$ (f) gh (g) $f + gh$ (h) $\dfrac{fg}{h}$

5 Work out the value of these expressions when $p = {}^-2$, $q = 4$ and $r = {}^-8$.

    (a) $2p + q$ (b) p^2 (c) $p(q + r)$ (d) $\dfrac{q}{p}$

    (e) $\dfrac{r}{p} + q$ (f) qr (g) ^-5p (h) $pq - r$

6 To work out roughly the surface area of a regular tetrahedron you can use the formula

$$A = \frac{7 \times e \times e}{4}$$

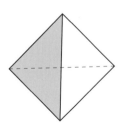

where A is the area in cm^2 and e is the length of an edge in cm.

Work out roughly the surface area of a regular tetrahedron with an edge of 12 cm.

Sections C and D

1

A	**B**	**C**
$17 + (a + b)$	$20 - (b + c)$	$20 - (a + c)$

When $a = 3$, $b = 4$ and $c = 5$, which of the expressions above is biggest?

2

A	**B**	**C**
$30 + (2p - q)$	$30 + (p - 2r)$	$30 + (q - 2p)$

Which of the expressions above gives the biggest value when

(a) $p = 3$, $q = {}^-2$ and $r = 4$ (b) $p = {}^-2$, $q = 5$ and $r = {}^-4$

3 Work out each of these when $x = {}^-2$, $y = 4$ and $z = {}^-3$.

(a) $20 - \dfrac{y}{x}$ (b) $20 - \dfrac{yz}{x}$ (c) $20 - \dfrac{y - x}{3}$

4 Work out the value of these when $a = 4$, $b = 6$, $c = \frac{1}{2}$ and $d = \frac{1}{3}$.

(a) $6 + \frac{1}{2}(b + a)$ (b) $10 - 6d$ (c) $12 - 2(a + c)$

5 James can use this formula to calculate the distance he is from home on his cycle ride home from Preston.

$$d = 60 - 12t$$

d is the distance in kilometres and t the time taken in hours.
He leaves Preston at 09:00.

(a) Find the value of d when $t = 4$.

(b) How far is he from home after a quarter of an hour?

(c) How far is he from home at half past nine?

(d) How many hours does it take James to reach home?

6 A formula for the width (w) of a rectangle is

$$w = \frac{p}{2} - l \qquad \text{where } p \text{ is the perimeter}$$
$$\text{and } l \text{ is the length of the rectangle.}$$

(a) What is the value of w when $p = 24.6$ and $l = 3.1$?

(b) What is the width of a rectangle with a length of 5.2 cm and a perimeter of 20.8 cm?

Section E

1 For the formula $v = 40 - 4t$, find

    (a) v when $t = 5$ (b) t when $v = 16$

2 This formula gives the cost of hiring a minibus (C) in pence.

        $C = 2000 + 20m$ where m is the number of miles travelled

    (a) What is the value of C when $m = 5$?

    (b) What is the value of m when $C = 4000$?

    (c) What is the cost if you travel 15 miles?

    (d) Bosh Street School hire a minibus.
        It costs them £28.

        How far did they travel?

3 This diagram shows a fence of
length a metres and height h metres.

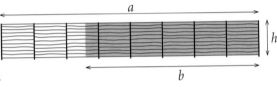

    One side of the fence has to be painted.
    The length of the part that has already been
    painted is b metres.

    (a) Write an expression for the area of the
        fence that still has to be painted.

    (b) If the length of the fence is 25 metres, the height is
        3 metres and they have already painted 12 metres
        what area is left to paint?

    (c) What area is left to paint if $a = 42$, $h = 1.5$ and $b = 10.6$?

4 In this triangle the length, in centimetres,
of the shortest side is x.

    One side is 4 cm longer than the shortest side.
    The other side is twice as long as the shortest side.

    (a) Write an expression, in terms of x,
        for the length of each side.

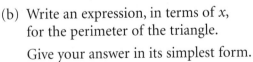

    (b) Write an expression, in terms of x,
        for the perimeter of the triangle.

        Give your answer in its simplest form.

    (c) The perimeter of the triangle is 24 cm.
        Use your answer to (b) to write down an equation.

    (d) Solve your equation and write down the lengths of the three sides.

Mixed questions 7

1 Work out the size of each of the angles marked with a letter.

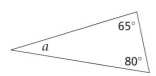   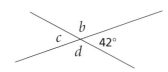   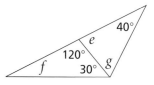

2 This box of cereal is in the shape of a cuboid.

   (a) Calculate the volume of the box.

   (b) Calculate the surface area of the box.

   Include the correct units in your answers.

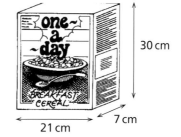

3 A sequence begins 2   8   14   20 ...

   (a) Write down the eighth term of the sequence.

   (b) Explain how you worked out your answer.

4 A birthday cake has a radius of 10 cm.
   What length of ribbon is needed to go around the outside of the cake?
   Give your answer to the nearest centimetre.

5 This table shows the distances in miles between some
   places in Scotland.

   (a) How far is it between Aberdeen and Glasgow?

   (b) Madge drives from Inverness to Aberdeen, and
        then from Aberdeen to Perth.
        How far has she driven altogether?

   (c) Gordon drove from Perth to Fort William.
        It took him $2\frac{1}{2}$ hours.
        What was his average speed?

Aberdeen				
157	Fort William			
147	102	Glasgow		
106	65	173	Inverness	
86	105	61	114	Perth

6 A group of year 10 students were asked how many hours they spent doing their
   homework last week.

   The results were: 12, 10, 8, 4, 3, 8, 9, 5, 8, 7, 11, 5

   (a) Work out the range of times spent doing homework.

   (b) Work out the median time spent.

   (c) Calculate the mean time.

7 Shape B is an enlargement of shape A.
What is the scale factor of the enlargement?

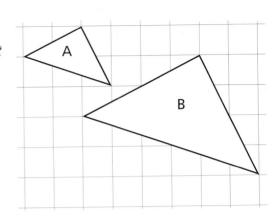

☒ 8 Work these out.

(a) $\frac{1}{4} + \frac{1}{3}$ (b) $\frac{2}{5} - \frac{1}{10}$ (c) $1\frac{1}{4} + 2\frac{1}{2}$ (d) $1\frac{3}{8} - \frac{3}{4}$

☒ 9 Which fraction is larger, $\frac{2}{3}$ or $\frac{5}{8}$?
Explain how you decided.

10 (a) Write down and simplify an expression for the perimeter of this pentagon in terms of x.

(b) The perimeter of the pentagon is 50 cm.
Find the value of x.

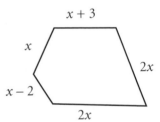

☒ 11 Work out the value of each of these when $x = 2$, $y = \frac{1}{2}$ and $z = \frac{1}{4}$.

(a) $x - y$ (b) $1 - (y + z)$ (c) $3x - 4z$ (d) $y - 2z$

12 Fraser went for a cycle ride.
His ride is represented by this graph.

(a) How far was he from home after 1 hour?
(b) What was his average speed between A and B?
(c) What happened between C and D?
(d) What was his average speed between D and E?
(e) How far did he cycle altogether?

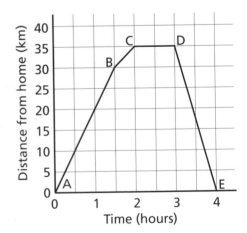

62 *Probability*

Sections A, B and C

1 Here is a probability scale.

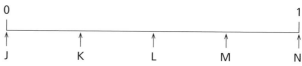

Which of the arrows could represent the probability of

(a) choosing a red counter from a bag with 3 blue counters and 9 red counters

(b) scoring a 7 on an ordinary dice

(c) choosing a red card from an ordinary pack of playing cards

2 In a game this fair spinner is used.

What is the probability it stops on

(a) the number 2   (b) an odd number

(c) a number less than 10   (d) the number 6

3 Louise carried out a survey on the students in her class. She asked them whether they would prefer a trip to Alton Towers or London. These are her results.

Trip	Alton Towers	London
Male	10	5
Female	6	9

(a) How many students were in Louise's class?

(b) If a student is chosen at random, what is the probability that

    (i) the student is female   (ii) the student wants to visit London

    (iii) the student wants to visit Alton Towers

4 Matt has a fair dice and coin.
He rolls the dice and flips the coin.

(a) Copy and complete this list of possible outcomes.

Dice	Coin
1	H
2	H

(b) What is the probability of getting

    (i) a 1 and a head   (ii) an even number and a tail

5 Jess has two fair spinners numbered as shown.
She spins them both and finds the total of the two scores.

(a) Copy and complete this grid to show all the possible total scores.

+	1	2	3	4	5	6
1						
2	3					
3			7			
4						

(b) Find the probability that the total score is

    (i) 10   (ii) 7

    (iii) less than 5   (iv) more than 1

Sections D and E

1 Lee flipped a coin and noted if it landed heads up.
   Convert his probabilities to decimals.

   (a) $\frac{7}{10}$　　　　　(b) $\frac{13}{20}$　　　　　(c) $\frac{23}{50}$　　　　　(d) $\frac{55}{100}$

2 Isaac is carrying out a survey about usage of his local library.
   He records the gender of the people entering the library.
   He writes male (M) or female (F) for each person.
   Here are his results in order:

   M F F M M F F F M　　　M F M M F F F M F M　　　F M F F F M M F M F

   M M F F F F F M F F　　　M F F M F F F M F F

   (a) Estimate the probability that a library user is female after

       (i)   the first 10 people have entered the library

       (ii) the first 20 people have entered the library

       (iii) the first 50 people have entered the library

   (b) The library staff have carried out their own survey, observing 500 library users.
        They say that the probability that a library user is female is 0.65.

        Do you think that Isaac's results agree with this?

3 Sally has made a dice with faces numbered 1 to 6.
   If the dice is fair the probability of scoring 6 is 0.17 to two decimal places.

   (a) Sally throws the dice 200 times and records the results.
        Copy and complete this table.

Score	1	2	3	4	5	6
Frequency	30	44	26	32	50	18
Estimated probability	0.15					

   $\frac{30}{200}$

   (b) Which numbers came up fewer times than expected?

   (c) Which numbers came up more times than expected?

   (d) Do you think that Sally's dice is biased?

Section F

1 Sean does a survey to see whether pupils in his school wear glasses or not.
After asking 100 pupils, he estimates the probability of a randomly chosen
student wearing glasses is 0.18.

   What is the probability of a randomly chosen student **not** wearing glasses?

2 A bag contains black and white beads.
The probability of choosing a black bead is 0.63.

   What is the probability of choosing a white bead?

3 Rosie has white, grey and pink socks in her drawer.
The probability of choosing a white sock is 0.5 and the probability of
choosing a grey sock is 0.4.

   (a) What is the probability of choosing a pink sock?

   (b) There are 40 socks in the drawer altogether.
        How many of these are pink?

4 Fraser has a large tub of building bricks. There are red, green and blue bricks.
He calculates that the probability of getting red is 0.48, blue 0.26 and green 0.24.

   Explain why these answers cannot be right.

5 Josie estimates that the probability she will be late for school is 0.24.

   (a) What is the probability that she is not late for school?

   (b) There are 52 school days in one term.
        On roughly how many days would Josie expect to be late?

6 Maninder has made this spinner.
He is testing to see whether it is biased.

   These are some of his results after a large number of trials.

Number	1	2	3	4	5
Probability	0.18	0.22	0.21	0.17	

   (a) What is the probability of getting a 5?

   (b) Maninder has spun the spinner 250 times.
        How many times did he get a 5?

   (c) Do you think Maninder's spinner is biased?
        Explain your answer.

63 *Getting more from your calculator*

Section B

1 Do each of these in your head. Then check each one using a calculator.
   (a) $3 \times 5 - 2$ (b) $3 + 2 \times 6$ (c) $20 - 3 \times 4$ (d) $5 + 5 \times 7$

2 Do each of these on a calculator.
   (a) $45 + 8 \times 16$ (b) $99 - 13 \times 6$ (c) $24 \times 37 + 201$ (d) $550 - 19 \times 21$

Do these on a calculator.

3 Calculate these. Round each answer to the nearest integer.
   (a) $8.6 \times 2.3 - 1.9$ (b) $15.2 + 4.8 \times 2.9$

4 Calculate these. Round each answer to one decimal place.
   (a) $34 + 0.7 \times 32.7$ (b) $46.5 - 3.7 \times 12.1$

5 Calculate these. Round each answer to two decimal places.
   (a) $15.75 + 2.44 \times 3.62$ (b) $0.89 \times 9.28 + 1.66$

Section C

1 Do each of these in your head. Then check each one using a calculator.
   (a) $5 + \dfrac{8}{2}$ (b) $\dfrac{15}{3} - 1$ (c) $9 + \dfrac{30}{5}$ (d) $16 - \dfrac{42}{7}$

2 Do each of these on a calculator.
   (a) $\dfrac{405}{45} + 11$ (b) $24 + \dfrac{544}{17}$ (c) $53 - \dfrac{966}{46}$ (d) $\dfrac{198}{12} - 14$

Do each of these on a calculator.

3 Calculate these. Round each answer to the nearest integer.
   (a) $3.15 + \dfrac{2.5}{0.6}$ (b) $84 - \dfrac{10.9}{1.7}$

4 Calculate these. Round each answer to one decimal place.
   (a) $42.4 + \dfrac{5.9}{1.2}$ (b) $\dfrac{3.7}{2.4} - 0.34$

5 Calculate these. Round each answer to two decimal places.
   (a) $\dfrac{15.6}{1.7} - \dfrac{2.4}{0.6}$ (b) $3.9 \times 5.1 - \dfrac{18.4}{3.1}$

6 Work out each of these, rounding to the nearest integer.
Put them in order, with the smallest first, to find a mathematical word.

M
$40.9 - 5.6 \times 3.25$

O
$\dfrac{51}{23} + \dfrac{235}{56}$

G
$\dfrac{2}{0.7} + \dfrac{3}{0.8}$

I
$5.5 - \dfrac{2}{1.5}$

L
$14.6 - 2.6 \times 3.8$

A
$8.2 + \dfrac{6.5}{3.9}$

R
$70.55 - 12.6 \times 4.88$

K
$\dfrac{41.2}{3.6} - 10.5$

Section D

1 Use a calculator to work these out.

(a) $^-12 + 23$ (b) $38 + {}^-27$ (c) $^-18 \times 23$ (d) $^-37 - {}^-48$

(e) $54 \div {}^-12.7$ (f) $^-4.2 \times {}^-3.7$ (g) $^-87 \div {}^-15$ (h) $8.5 + {}^-12.7$

(i) $8.5 \div {}^-1.7$

2 (a) Between 23rd and 24th January 1916, the temperature in Browning, Montana, in the USA fell from 6.7°C to ⁻48.9°C.
By how much did the temperature fall?

(b) At Rapid City, Dakota, on 10th January 1911, the temperature at 07:00 was 12.8°C but by 07:15 it had dropped to ⁻13.3°C.
By how much had the temperature dropped in that 15 minutes?

Sections E and F

1 Do these on a calculator. Round the answers to one decimal place.

(a) $(56 - 29) \times 0.24$ (b) $(3.12 + 2.45) \times 5.26$ (c) $3.14 \times (4.23 - 1.86)$

(d) $\dfrac{85 - 38}{34}$ (e) $\dfrac{45.6 + 21.9}{8.34}$ (f) $\dfrac{8.23 \times 2.6}{2.5}$

2 Do these on a calculator.

(a) $10 + ({}^-3 - 5)$ (b) $7 - ({}^-3 + 2)$ (c) $^-2.5 \times (3.4 - 1.2)$

3 Calculate these, giving each answer correct to two decimal places.

(a) 6.25^2 (b) $6.99^2 + 1.2^2$ (c) $(3.6 - 1.25)^2$ (d) $4.67 + 1.35^2$

4 Calculate each of these. Round each answer to two decimal places.

(a) 2.8^3 (b) 5.1^4 (c) 3.5×1.2^3 (d) $2.6 - 1.2^4$

Section G

1 Use a calculator to find

   (a) $\sqrt{676}$ (b) $\sqrt{67.24}$ (c) $\sqrt{0.4489}$

2 Use a calculator to evaluate these.

   (a) $4 - \sqrt{8.41}$ (b) $7.2 \times \sqrt{39.69}$ (c) $\dfrac{\sqrt{17.64}}{3}$ (d) $3.8^2 - \sqrt{1.96}$

3 Evaluate each of these expressions.

   (a) $400 + \sqrt{441}$ (b) $\sqrt{400 + 441}$ (c) $\sqrt{361} - \sqrt{225}$ (d) $\dfrac{\sqrt{27.04}}{\sqrt{1.69}}$

4 Work these out, giving your answers correct to two decimal places.

   (a) $4.5^2 - \sqrt{12.3}$ (b) $\sqrt{8.2 + 3.7} - 2.5$ (c) $1.72 \times \sqrt{2.4 + 5.3}$

5 Work out each of these, rounding to two decimal places.
   Put them in order, with the smallest first, to find a mathematical word.

M	I	R	C	E
$\dfrac{4.05^2}{2.3}$	$\dfrac{\sqrt{42.25}}{1.45}$	$\dfrac{(14 + 8.75)^2}{6.5}$	1.04^2	3.2×2.13^2

T	N	E	E	T
$\sqrt{284 - 73}$	$\sqrt{0.88 - 0.23} + 1.1^2$	1.08^3	$4.3^3 + \dfrac{6.5}{50}$	$\dfrac{8.25 + \sqrt{9.61}}{2.6}$

Section H

1 What are the reciprocals of these numbers?

   (a) 5   (b) 16   (c) 40   (d) 200
   (e) 0.02   (f) 0.25   (g) 0.0125   (h) 3.2

2 Use a calculator to work these out.

   (a) $\dfrac{1}{8}$ (b) $\dfrac{1}{20}$ (c) $\dfrac{1}{6.4}$ (d) $\dfrac{1}{250}$

3 Use a calculator to work these out.

   (a) $\dfrac{1}{8.2 - 6.6}$ (b) $9.2 + \dfrac{1}{8}$ (c) $\dfrac{1}{\sqrt{6.25}}$

64 Transformations

Sections B, C and D

1. (a) Draw a grid with both axes going from ⁻5 to 5.
   (b) Draw the shape with coordinates (3, 2) (5, 2) (5, 3) (4, 3) and (3, 4). Label it A.
   (c) Reflect shape A in the line x = 1. Label it B.
   (d) Reflect shape B in the x-axis. Label it C.
   (e) Reflect shape A in the line y = 0. Label it D.
   (f) What line would be the mirror line to map shape C on to shape D?

2. Describe the translations that map
   (a) shape 1 on to shape 2
   (b) shape 3 on to shape 2
   (c) shape 1 on to shape 4
   (d) shape 3 on to shape 4

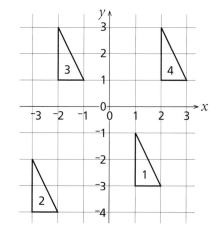

3. Where would the point (3, ⁻2) go to after these translations?
   (a) 5 right and 3 down
   (b) 4 left and 4 up

4. Describe the transformation on this diagram that maps
   (a) shape A on to shape B
   (b) shape D on to shape B
   (c) shape C on to shape B

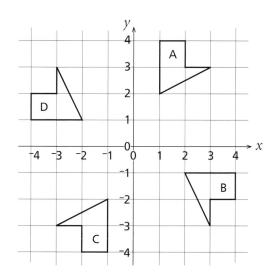

Sections E and F

1 (a) Copy this graph and shape on to squared paper.

(b) Enlarge the shape by scale factor 3 with centre (0, 0).

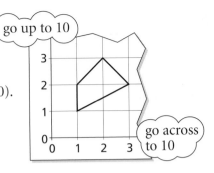

2

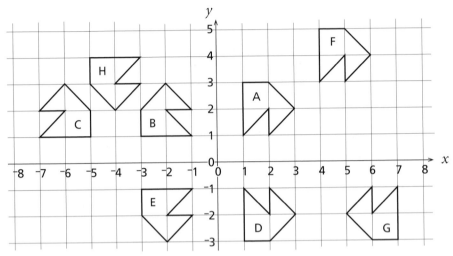

Describe fully the transformation that will map

(a) shape A on to shape B

(b) shape B on to shape C

(c) shape A on to shape D

(d) shape D on to shape E

(e) shape A on to shape F

(f) shape D on to shape G

(g) shape E on to shape H

3 (a) Draw a grid with both axes going from ‾4 to 4.

(b) Draw the trapezium with coordinates (2, 1) (4, 1) (4, 2) and (3, 2). Label it A.

(c) Reflect shape A in the x-axis. Label it B.

(d) Rotate shape B 90° clockwise, centre (0, 0). Label it C.

(e) Translate shape C by 1 left and 6 up. Label it D.

(f) Rotate shape C 180° centre (0, 0). Label it E.

(g) Draw the mirror line that will reflect shape E on to shape A.

65 *Written calculations with decimals*

Section B

1 Use the fact that $23 \times 47 = 1081$ to write down the answers to

   (a) 2.3×4.7 (b) 0.23×4.7 (c) 23×0.47 (d) 0.23×0.47 (e) 2.3×0.47

2 Work out a rough estimate for each of these.

   (a) 4.9×5.1 (b) 0.71×5.2 (c) 31×0.82 (d) 59×0.19 (e) 0.23×6.8

3 Work these out, using your answers to question 2 to check that your answers are reasonable.

   (a) 4.9×5.1 (b) 0.71×5.2 (c) 31×0.82 (d) 59×0.19 (e) 0.23×6.8

4 Monty bought 8.5 litres of petrol. It cost 82p per litre.

   How much did Monty spend on petrol?

5 Kylie bought 3.5 square metres of carpet.
   It cost £8.70 per square metre.

   How much did Kylie spend on carpet?

6 A house has three bedrooms. Their dimensions are

   (a) 2.2 m by 3.1 m   (b) 3.3 m by 2.9 m   (c) 4.2 m by 3.6 m

   Work out the floor area of each bedroom.

7 Gemma was investigating the sizes of photographs in newspapers for a data handling project. The first five she looked at had these dimensions

   (a) 21 cm by 16 cm   (b) 7.2 cm by 12 cm   (c) 3.9 cm by 5.4 cm

   (d) 8.4 cm by 8.4 cm   (e) 4.5 cm by 5.1 cm

   Find the area of each photograph.

8 Rewrite this paragraph with all the imperial measurements converted to metric measurements.

   'Jennifer is 5 foot tall. She weighs 98 pounds. Her handspan is $5\frac{1}{2}$ inches.
   On average, she drinks 1.4 pints of water a day. She lives 0.7 miles from her school.'

1 foot is 0.3 m	1 pound is 0.45 kg	1 inch is 2.5 cm
1 pint is 0.57 litres		1 mile is 1.6 km

Section C

1 Paul went to the supermarket and bought
0.5 kg of carrots, 0.4 kg of tomatoes and 1.2 kg of turnips.

Carrot	£0.58 per kilo
Tomato	£1.80 per kilo
Turnip	£0.85 per kilo

(a) Work out the cost of each vegetable.

(b) Paul paid with a £20 note.
How much change did he receive?

2 Ron used five pieces of wood each 0.27 m long.

(a) Find the total length of wood he used.

He cut the pieces of wood from a plank which was 1.8 m long.

(b) How much wood did he have left over?

3 A 'stone' is an imperial weight measurement.
1 stone is equivalent to 14 pounds.

(a) Convert 1 stone to kilograms (1 pound is 0.45 kg).

(b) Michelle weighs 9 stone. What is her weight in kilograms?

4 A cake stall at a summer fête sold chocolate cakes for £0.35 each,
lemon tarts for £0.25 each and rock buns for £0.15 each.
They sold 54 chocolate cakes, 26 lemon tarts and 42 rock buns.

(a) How much was spent on

    (i) chocolate cakes   (ii) lemon tarts   (iii) rock buns

(b) How much money was spent altogether?

5 A floor is tiled with two different square tiles.
Here is part of the pattern made with the tiles.

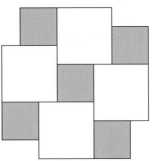

(a) The white tiles measure 0.3 m by 0.3 m.
What is the area of a white tile?

(b) The coloured tiles measure 0.2 m by 0.2 m.
What is the area of a coloured tile?

(c) What is the total area of a white tile
and a coloured tile?

(d) 80 white tiles and 80 coloured tiles are used to cover a floor.
What is the area of the floor covered?

(e) 9 white tiles and 9 coloured tiles fit along one wall of a room.
What is the **length** of the wall?

66 *The solution is clear*

Sections A and B

1 Write an equation for each puzzle.
Solve it to find the weight of a can.

(a)

(b)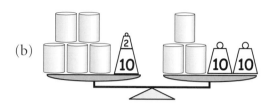

2 Solve each equation and check your answers.

(a) $3x + 7 = 19$ (b) $6x - 3 = 9$ (c) $13 = 4x + 1$

(d) $5p - 4 = 3p + 2$ (e) $b + 12 = 3b - 8$ (f) $7f + 3 = 8f + 1$

3 The angles in a quadrilateral add up to 360°.
Use this fact to write an equation involving p for each quadrilateral.

Solve each equation.

(a)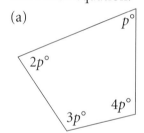

(b)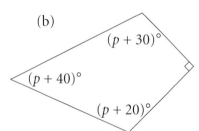

4 Solve each of these equations.

(a) $4n + 3 = 17 - 3n$ (b) $3p - 7 = 21 - p$ (c) $2n - 5 = 3 - 2n$

(d) $20 - 5n = 3n + 4$ (e) $25 - 7p = p + 1$ (f) $6x + 7 = 15 - 2x$

5 Solve

(a) $3 = 12 - 3x$ (b) $5 = 17 - 4x$ (c) $28 - 7x = 7$

6 Work out what x stands for in each of these.

(a)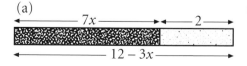

(b)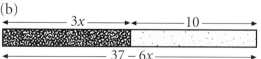

7 Solve these. The answers may be negative or include fractions.

(a) $5x + 9 = 4$ (b) $12 - n = 15$ (c) $3x - 4 = {}^{-}2$

(d) $3 - d = 3d$ (e) $x = 9 - x$ (f) $10 - 3x = 6$

8 Solve

   (a) $n + 10 = 19 - 2n$ (b) $4n + 1 = 11 - n$

   (c) $2p + 3 = 10 - 5p$ (d) $3p + 7 = 7 - 2p$

Sections C and D

1 Solve each of these equations.

   (a) $4(x + 2) = 32$ (b) $5(x - 3) = 30$ (c) $3(n - 4) = 27$

   (d) $4(x - 5) = 2x$ (e) $2(u + 12) = 5u$ (f) $y = 4(y - 3)$

2 Solve each of these.

   (a) $2(3 - x) = 4x$ (b) $9 = 3(4 - x)$ (c) $4(3x - 1) = 20$

   (d) $5(2y + 7) = 45$ (e) $3(7x - 4) = 30$ (f) $2(9x - 1) = 7$

3 Solve

   (a) $2(x + 5) = x + 14$ (b) $3(n - 3) = n + 1$ (c) $4(x - 3) = 8(x - 5)$

4 By multiplying to get rid of the fractions, solve each of these equations.

   (a) $\dfrac{3x}{4} = 9$ (b) $\dfrac{y + 3}{4} = 7$ (c) $\dfrac{3p - 2}{2} = 5$

5 (a) Form an equation involving x for each of the following shapes.

   (b) Solve each equation.

   (c) Sketch each shape, showing the values of the angles.

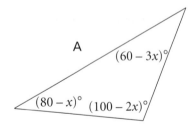

6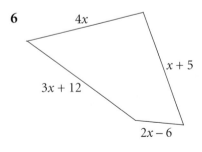

   (a) The lengths shown here are in centimetres. Write down an expression in x for the perimeter of this quadrilateral. Give your answer in its simplest form.

   The perimeter is 51cm.

   (b) Write down an equation in x and solve it.

   (c) Write down the lengths of the four sides.

67 *Multiplying and dividing fractions*

Sections B and C

1 Five bars of chocolate are shared equally between 4 people.
How much does each person get?

2 Work these out writing your answers as mixed numbers.

(a) $\frac{1}{3}$ of 5   (b) $\frac{1}{4}$ of 15   (c) $\frac{1}{5}$ of 7   (d) $\frac{1}{6}$ of 13

3 Work these out giving your answers as mixed numbers in their simplest form.

(a) $\frac{1}{4}$ of 22   (b) $\frac{1}{8}$ of 20   (c) $\frac{1}{6}$ of 15   (d) $\frac{1}{3}$ of 11

4 Work out each answer and use the code to change it to a letter.
(You may need to simplify your answer.)

Then rearrange each set of letters to spell a piece of furniture.

S	O	B	H	K	L	T	E	A	C	N	D
$\frac{2}{3}$	$\frac{3}{4}$	$1\frac{1}{2}$	$1\frac{2}{5}$	$1\frac{3}{5}$	$1\frac{4}{5}$	$2\frac{1}{3}$	$2\frac{1}{2}$	$2\frac{2}{3}$	$2\frac{3}{4}$	$3\frac{3}{4}$	$4\frac{3}{4}$

(a) $\frac{1}{4}$ of 10 $\frac{1}{8}$ of 12 $\frac{1}{4}$ of 19

(b) $\frac{1}{10}$ of 18 $\frac{1}{3}$ of 8 $\frac{1}{2}$ of 5 $\frac{1}{3}$ of 7 $\frac{1}{2}$ of 3

(c) $\frac{1}{6}$ of 14 $\frac{5}{6}$ of 3 $\frac{1}{8}$ of 22 $\frac{1}{5}$ of 7 $\frac{2}{9}$ of 3

5 Do the same as for question 4, but rearrange each set of letters to spell an item of food.

R	A	D	S	C	B	E	H	I	F	T	P
$\frac{6}{7}$	$1\frac{1}{5}$	$1\frac{1}{3}$	$1\frac{1}{2}$	$1\frac{2}{3}$	$2\frac{1}{4}$	$2\frac{2}{5}$	$2\frac{4}{5}$	$3\frac{1}{5}$	$3\frac{1}{3}$	$3\frac{3}{4}$	$6\frac{3}{4}$

(a) $\frac{5}{6}$ of 2 $\frac{2}{7}$ of 3 $\frac{4}{5}$ of 4 $\frac{3}{5}$ of 4

(b) $\frac{2}{5}$ of 6 $\frac{3}{8}$ of 6 $\frac{2}{3}$ of 5 $\frac{4}{5}$ of 3

(c) $\frac{3}{10}$ of 5 $\frac{4}{5}$ of 4 $\frac{3}{8}$ of 18 $\frac{5}{9}$ of 3 $\frac{2}{5}$ of 7

6 This piece of wood is 12 inches long.
It is divided into 5 equal sections.

(a) How many inches long is each section?
Write your answer as a mixed number.

(b) How long is the distance from A to B on the piece of wood?

Sections D and E

1 Work these out.

    (a) $\frac{1}{4} \div 3$ (b) $\frac{1}{5} \div 2$ (c) $\frac{1}{2} \div 5$

2 Matthew has $\frac{1}{3}$ of a cake.

   He shares it equally with his brother and sister.

   What fraction of the cake do they each get?

3 Work these out.

    (a) $\frac{1}{3}$ of $\frac{1}{4}$ (b) $\frac{1}{4}$ of $\frac{1}{5}$ (c) $\frac{1}{5}$ of $\frac{1}{2}$

4 Daniel has $\frac{3}{4}$ of a cake. He shares it equally with his friend Heather.

   What fraction of the cake do they each get?

5 Work these out.

    (a) $\frac{2}{5} \div 3$ (b) $\frac{2}{3} \div 4$ (c) $\frac{3}{4} \div 5$

    (d) $\frac{1}{2}$ of $\frac{2}{5}$ (e) $\frac{1}{4}$ of $\frac{3}{8}$ (f) $\frac{1}{5}$ of $\frac{2}{3}$

Section F

1 (a) $\frac{1}{2} \times \frac{1}{4}$ (b) $\frac{2}{5} \times \frac{1}{4}$ (c) $\frac{3}{4} \times \frac{1}{2}$

2 Work out the answers to each of these.

    (a) $\frac{3}{5} \div 2$ (b) $\frac{1}{3}$ of $\frac{2}{5}$ (c) $\frac{1}{2}$ of $1\frac{1}{3}$ (d) $\frac{1}{3} \times \frac{2}{7}$

    (e) $\frac{1}{8}$ of $\frac{3}{4}$ (f) $\frac{5}{8} \div 3$ (g) $\frac{1}{4} \times \frac{3}{4}$ (h) $\frac{1}{5}$ of $2\frac{1}{2}$

3 Put these calculations into pairs which give the same answer.

   Write down the answer to each pair.

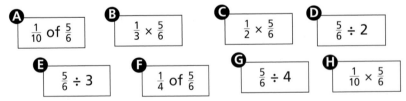

A $\frac{1}{10}$ of $\frac{5}{6}$ **B** $\frac{1}{3} \times \frac{5}{6}$ **C** $\frac{1}{2} \times \frac{5}{6}$ **D** $\frac{5}{6} \div 2$

E $\frac{5}{6} \div 3$ **F** $\frac{1}{4}$ of $\frac{5}{6}$ **G** $\frac{5}{6} \div 4$ **H** $\frac{1}{10} \times \frac{5}{6}$

68 Constructions

Sections A, B, C and D

1 Draw these triangles accurately, then measure the angle and lengths marked with letters.

(a)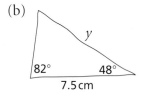

(b)

(c)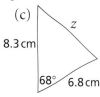

2 (a) Draw accurately triangle ABC where AB = 9 cm, AC = 5.4 cm, BC = 7.2 cm.

(b) Measure angle ACB.

(c) What type of triangle is ABC?

(d) Calculate the area of triangle ABC.

3 (a) Make an accurate copy of this quadrilateral.

(b) What special type of quadrilateral is it?

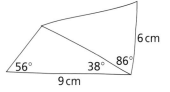

4 One of these triangles can be drawn in two different ways. From these sketches draw three accurate triangles.

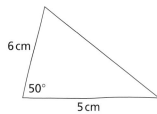
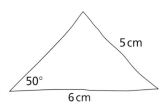

5 (a) Draw a line BC, 5 cm long.
Complete triangle ABC with angle B = 70° and BA = 5 cm.

(b) Use AC as the base of another triangle ADC where AD = CD = 5 cm and D is on the opposite side of AC to B.

(c) What type of quadrilateral is ABCD?

6 (a) Draw an accurate plan of this park, using a scale of 1 cm to 10 m.

(b) There are two straight paths in the park, one from Main Gate to Honeysuckle Gate and the other from Archway to Dewdrop Gap.

What is the length, in metres, of the longer path?

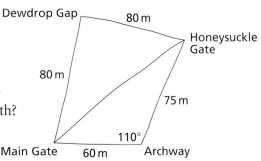

69 Percentage increase and decrease

Sections B, C and D

1 A bottle contains 330 ml of lemonade. Deborah drinks 10% of the lemonade.

   (a) How much lemonade did Deborah drink?

   (b) How much lemonade is left in the bottle?

2 Mark earned £80 last week. This week his earnings increased by 10%.

   (a) What is 10% of £80?

   (b) How much did Mark earn this week?

3 In a sale, a shop reduced its prices by 5%. A television cost £300 before the sale.

   (a) What is 5% of £300?

   (b) How much did the television cost in the sale?

4 Two sunflower plants were entered into a 'Tallest Sunflower' competition.
   Errol's sunflower plant was 3 m 20 cm tall but Gwyneth's was 5% taller.

   (a) What is 5% of 3 m 20 cm?

   (b) How tall was Gwyneth's sunflower plant?

5 Last season, the local football club had 4800 season ticket holders.
   The number of season ticket holders increased by 25% this season.

   (a) How many extra season tickets were sold this season?

   (b) How many season tickets were sold this season?

6 A car cost £9000 when it was new. Its value has decreased by 20%.
   What is the car worth now?

7 Ten years ago, a house was worth £120 000.
   Since then, its value has increased by 50%.
   What is the house worth now?

8 Last year, 80 students at one school achieved a grade C for mathematics GCSE.
   This year, the number of students achieving a C increased by a quarter.
   How many students achieved a grade C this year?

9 It took Puja 12 hours to drive to her holiday chalet.
   Her journey time for the return journey was a third less.
   How long did it take Puja to drive home?

Sections E and F

1 Calculate

   (a) 37% of £86   (b) 43% of £420   (c) 8% of 350 g   (d) 81% of 93 km

2 Calculate the following, rounding your answers to the nearest penny.

   (a) 11% of £8.45   (b) 3% of £6.85   (c) 97% of £13.68

3 27% of the children in a town attend Marden High School.
  There are 5300 children in the town.
  How many children attend Marden High School?

4 3400 people took part in a survey on the back of a cereal packet.

   (a) 38% drank a cup of tea with their breakfast.
      How many drank tea?

   (b) 8% ate some fruit with their cereal.
      How many people ate some fruit?

   (c) 72% ate a piece of toast.
      How many people ate a piece of toast?

5 A factory employs 350 workers.
  The number of workers is increased by 6%.
  How many workers does the factory employ now?

6 VAT at the rate of 17.5% is added to the basic cost of a camera.
  The basic cost of a camera is £170.20.

   (a) To the nearest penny, how much is the VAT?

   (b) What is the cost of the camera, including the VAT?

7 A savings account has an interest rate of 4.3% per annum.
  A year ago, Sanjay put £240 into the account.

   (a) How much interest has Sanjay's money earned?

   (b) Has Sanjay got enough money in his account to pay for an aeroplane flight costing
      £250?

8 Hayley's mum bought a new washing machine.
  The washing machine cost £380.
  Hayley's mum paid a deposit of 15% and then 6 monthly payments of £58.

   (a) How much deposit did Hayley's mum pay?

   (b) How much altogether did Hayley's mum pay for the washing machine?

   (c) How much would she pay if she had paid a deposit of 25% and 12 monthly
      payments of £27?

70 Conversion graphs

Sections A and B

1 The areas of floors and carpets may be measured in square metres (m^2) or square yards.

This graph can be used to convert between square yards and m^2.

Use the graph to convert

(a) 15 square yards to m^2

(b) 22 square yards to m^2

(c) 20 m^2 to square yards

(d) 9 m^2 to square yards

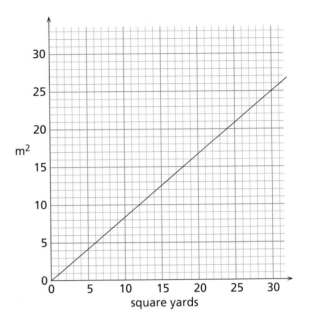

2 In older cookery books, liquids may be measured in 'fluid ounces'.
This table shows volumes in fluid ounces and in millilitres (ml).

Fluid ounces	0	5	10	15	20	25
Millilitres	0	140	280	420	560	700

(a) On graph paper draw axes as shown here. Draw a conversion graph.

(b) Use your graph to convert

(i) 18 fluid ounces to ml

(ii) 250 ml to fluid ounces

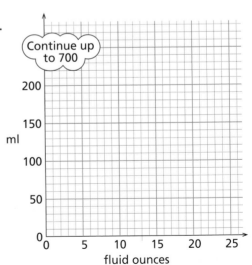

Continue up to 700

Mixed questions 8

1   Kyle has a large bag of sweets containing mints, toffees and fruit gums.
    When he picks a sweet at random, the probability that it is a mint is 0.25 and the
    probability that it is a toffee is 0.4.

    (a) What is the probability that it is a fruit gum?

    (b) There are 80 sweets in the bag.
         How many of them are toffees?

2   Maya makes a spinner numbered from 1 to 5.
    To test the spinner she spins it 200 times and records the results.

    (a) Copy and complete this table of Maya's results.

Number	1	2	3	4	5
Frequency	38	42	26	51	43
Estimated probability	0.19				

    (b) Is Maya's spinner biased?
         Explain why you think this.

3   Use a calculator to evaluate these expressions.

    (a) $12 + \sqrt{1156}$ (b) $\dfrac{\sqrt{5.76}}{4}$ (c) $2.2 \times \sqrt{4.36 + 2.93}$

4   Draw a grid on squared paper
    with both axes going from ⁻5 to 5.
    Copy shape P on to your grid.

    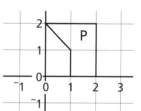

    (a) Draw the reflection of shape P in the line $y = ^-1$.
         Label this Q.

    (b) Draw the enlargement of shape P, scale factor 2
         with centre $(0, 0)$. Label this R.

    (c) Draw the image of shape P after a 180° rotation,
         centre $(0, 0)$. Label this S.

5   (a) Work out a rough estimate for 0.24×150.

    (b) Work out the exact answer to 0.24×150.

6   1 mile is equivalent to 1.6 km.
    Yasmin completes a 25 mile cycle ride.

    How many kilometres is this?

7   Matt is buying pizzas for a children's party.
    He needs $\frac{1}{3}$ of a pizza for each child.

    How many pizzas will he need for 12 children?

8 (a) Write down and simplify an expression for the sum of the interior angles in this quadrilateral, in degrees.

(b) By forming an equation, find the value of x.

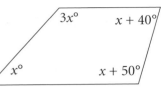

9 Triangle ABC has sides of length AB = 6 cm, BC = 8 cm and AC = 4 cm.

(a) Using a ruler and compasses only, make an accurate drawing of the triangle.

(b) Measure and write down the size of the obtuse angle in the triangle.

10 Solve these equations.

(a) $5x - 2 = 4x + 9$ (b) $3(y + 2) = 21$ (c) $12 - 4n = 2n$

11 Work out the cost of each of these.

(a) 1 kg of almonds and 1 kg of hazelnuts

(b) 0.6 kg of Brazil nuts

(c) 1.8 kg of walnuts

(d) 0.8 kg of almonds and 2 kg of hazelnuts

12 A DVD player originally cost £250. Its price is reduced by 20% in a sale. What is the sale price of the DVD player?

13 Rashid invested £480 in a savings account with an interest rate of 3.5% per annum.

(a) How much interest will his investment have earned after one year?

(b) How much in total will his investment now be worth?

14 This graph can be used to convert between inches and centimetres.

(a) Use the graph to convert

(i) 2 inches to centimetres

(ii) 9 centimetres to inches

(b) Use the graph to help you convert

(i) 10 inches to centimetres

(ii) 15 centimetres to inches

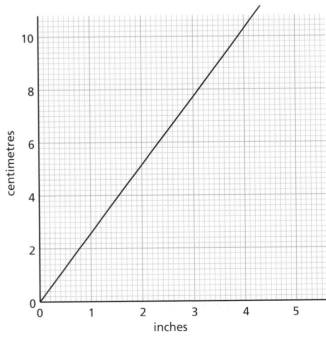

General review

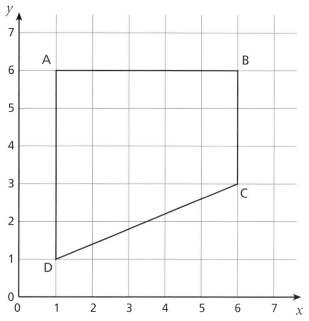

1 (a) What is the name of shape ABCD?

(b) What are the coordinates of point A?

(c) Measure the length of CD in centimetres.

(d) Find the perimeter of ABCD.

(e) Work out the area of ABCD.

2 (a) Copy and complete this table with the correct equivalents.

Fraction	Decimal	Percentage
$\frac{1}{4}$		
	0.1	
		30%

(b) What fraction of this rectangle is shaded? Write your answer in the simplest form.

(c) Work out $\frac{1}{6} + \frac{1}{2}$. Simplify your answer.

3 Work these out.

(a) $16 - 4 \times 3$ (b) 18×23 (c) $224 \div 14$

4 The table shows the temperature on January 1st at noon in 5 cities.

London	Moscow	Sydney	Oslo	Bombay
1°C	⁻8°C	12°C	⁻5°C	20°C

(a) Which city had the highest temperature?

(b) Which city had the lowest temperature?

(c) Sydney was hotter than Oslo. By how many degrees?

(d) By 6 p.m., the temperature in Moscow had dropped 5 degrees. What was the temperature in Moscow at 6 p.m.?

5 Look at the numbers in this box.

2	12	15	16	25	27

    (a) Which of the numbers are even?

    (b) One of the numbers is a multiple of 6.
        Which is it?

    (c) One number is prime. Which one?

    (d) Which of the numbers are square numbers?

6 (a) Simplify $x + x + x$.   (b) Expand $3(x - 2)$.   (c) Factorise $5x + 15$.

7 Write these numbers in order, starting with the smallest.

        0.2   0.19   0.09   0.1   0.099

8 Choose the unit from the box that would be most appropriate to use when measuring

| grams |
| kilograms |
| tonnes |
| millilitres |
| litres |
| cubic metres |

    (a) the weight of a person

    (b) the volume inside a furniture van

    (c) the weight of a train

    (d) the volume of a tropical fish tank

    (e) the weight of a teaspoon of flour

    (f) the volume of a teaspoon of water

9 Look at these patterns.

    (a) How many dots will there be in pattern 4?

Pattern 1   Pattern 2   Pattern 3

    (b) Copy and complete this table.

Pattern number	1	2	3	4	5
Number of dots					

    (c) (i) Work out how many dots there will be in pattern 30.

        (ii) Explain how you worked out your answer.

10 (a) A box contains 5 mints and 10 fruit flavoured sweets.
       I pick a sweet at random. What is the probability it is a mint?

    (b) A class survey of some students gave the results in the table.

	Boys	Girls
Left-handed	3	2
Right-handed	14	11

       (i) How many students are in the class?

       (ii) A student is chosen at random from the class.
          What is the probability that the student is right-handed?

11 (a) Nita wants to buy some 28p stamps.
How many can she buy with £5?

(b) James also wants some 28p stamps.
He buys 14 stamps and pays with a £10 note.
How much change does he get?

(c) Rachel buys some 28p stamps and pays with a £5 note.
She gets £1.64 change.
How many stamps does she buy?

12 (a) A sequence starts   2   3   5   9 ...

The rule for the sequence is   Double the last number and take off 1.

What is the next number in the sequence?

(b) A sequence with a different rule starts   ⁻3   ⁻5   ⁻7   ⁻9 ...

(i) What is the rule for this sequence?

(ii) Write down the next two numbers in this sequence.

13 This newspaper headline is about a march.

35 600 march to say No!

(a) What is 35 600 to the nearest thousand?

(b) 45% of the marchers were female.
How many female marchers were there?

14 *Auto Savers* hire out cars.
There is a basic charge of £24.50 per day, plus 12p per mile.

Anita hires a car for two days.

This was her mileometer reading at the start.   This is the mileometer at the end.

1 4 5 0 6		1 4 7 3 8

Work out the total hire charge that Anita has to pay.

15 Answer these questions, giving your answers to a sensible degree of accuracy.
Prestige Tables make circular wooden tables.

(a) The *Economy* table has a diameter of 1.2 metres.
What is the circumference of the *Economy* table?

(b) The *Standard* table has a diameter of 1.5 metres.
What is the area of the *Standard* table?

(c) A new table design is to have an area of $2\,m^2$.
What diameter should this table have?

16 Work these out, giving your answers to one decimal place.

(a) $\dfrac{4.8 - 3.92}{0.53}$ (b) $\dfrac{9.64}{2.7 + 1.03}$ (c) $\sqrt{2.35^2 + 3.2^2}$

17 In the Corner Café, a coffee costs 90p and a cup of tea 65p.

   (a) What is the cost of four coffees and three teas?

   (b) Oliver buys n cups of tea.
      Write an expression for the cost, in pence.

   (c) Ditta buys x coffees and y teas.
      Write an expression for the total cost, in pence.

18 If $m = 5$, $n = 6$ and $p = \frac{1}{2}$, work out

   (a) $4m + 5n$ (b) $2(n + p)$ (c) $p(3m + 4n)$

19 (a) Write down an expression in terms of x
        for the sum of the angles of this triangle.

        Give your answer in its simplest form.

   (b) Write down an equation in terms of x.

   (c) By solving your equation, work out the
        size of each angle in the triangle.

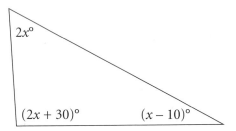

20 Ann and Beth are high-jumpers.
    They keep a record of their last few jumps, in metres.

    Ann   1.56 1.49 1.65 1.48 1.57
    Beth   1.87 1.34 1.81 1.38

   (a) Copy and complete this table.

   (b) Which high-jumper would you pick
        for a competition.
        Give two reasons for your choice.

	Mean	Range
Ann	1.55 m	0.17 m
Beth		

21 The diagram shows the net of a cuboid.

   (a) Make a rough sketch of the diagram.

   (b) On your diagram, write down the
        lengths shown by letters.

   (c) Work out the volume of the cuboid.

   (d) Work out the surface area of the cuboid.

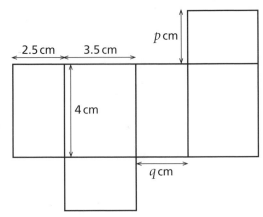